Matrix
Representations
of
Groups

MORRIS NEWMAN

DOVER PUBLICATIONS, INC., Mineola, New York

Bibliographical Note

This Dover edition, first published in 2019, is an unabridged and newly reset republication of the work originally published in 1968 as Volume 60 in the Applied Mathematics Series issued by the National Bureau of Standards, Washington, D.C.

Library of Congress Cataloging-in-Publication Data

Names: Newman, Morris, 1924–2007, author.
Title: Matrix representations of groups / Morris Newman.
Description: Dover edition. | Mineola, New York : Dover Publications, Inc.,
    2019. | Originally published: Washington, D.C. : National Bureau of Standards, 1968
    (Applied mathematics series ; volume 60). | Includes bibliographical references.
Identifiers: LCCN 2018059409 | ISBN 9780486832456 | ISBN 0486832457
Subjects: LCSH: Representations of groups. | Matrices.
Classification: LCC QA176 .N49 2019 | DDC 512/.22—dc23
LC record available at https://lccn.loc.gov/2018059409

Manufactured in the United States by LSC Communications
83245701   2019
www.doverpublications.com

Contents

Foreword

The theory of group representations is of fundamental importance in such disciplines as particle physics and crystallography and has been a major force in group theory. Thus the celebrated theorem of J. Thompson and W. Feit, that every group of odd order is solvable, depends heavily for its proof on this theory; and the new classifications of particles are described by means of certain special continuous groups and their representations.

For some time there has been a need for a simple but complete exposition of this subject and the present volume should meet this need. It should be of value to the worker in the field who has occasion to use the subject or who must understand it. The many special representations worked out in detail should also prove quite useful.

A. V. Astin, Director
National Bureau of Standards

Introduction

In this volume the most important facts about group representations are developed, entirely along the original matrix-theoretic lines set down by Burnside, Frobenius, and Schur in their fundamental memoirs on this subject. In the writer's opinion the approach through matrices is the one most easily grasped by the beginning student and is also the one which is most easily applied to other parts of mathematics and other disciplines. The more general approach through modules is not discussed.

Very little is presupposed about groups, but the reader should certainly be familiar with classical matrix theory: a good reference source containing all necessary material is the tract by MacDuffee [24].[1] Another is the volume by Marcus [25] in this Series. Any nonstandard results from group theory or matrix theory are proved in the text.

An appendix on the elements of the theory of algebraic numbers has been included, so that the volume is self-contained in this respect. This appendix (which parallels the introductory material in Hecke's book [19] quite closely) serves by itself as a complete introduction to the classical theory of algebraic numbers, up to and including the unique factorization theorem for ideals. There is also an appendix on the roots of unity, containing an elementary proof of the irreducibility of the cyclotomic polynomial.

The entire development has arbitrarily been limited to the case when the ground field is the field of complex numbers (so that modular representations are not even discussed). The experienced reader will see that many of the proofs are valid for arbitrary fields, or can be modified slightly to be so. Furthermore many important topics, such as projective representations, are not mentioned. On the other hand all of the important elementary results have been included, a number of advanced

---

[1] Figures in brackets indicate the literature references at the end of this paper.

topics are treated, and a considerable number of special representations have been worked out in detail.

The volume follows in spirit the lectures given by Schur and prepared by Stiefel at Zurich in 1936 [27]. Also the original papers of Burnside, Frobenius, and Schur were consulted frequently, and many of the discussions follow these papers quite closely.

Without doubt the most important modern book on this subject is the book by Curtis and Reiner [9], and this has been consulted as well. The ultimate selection of subjects was of course a matter of taste and reflects the writer's personal likes and dislikes.

The writer's principal objective was to make available the results and techniques of the subject of group representations to an audience with at most a standard mathematical background, and to indicate some interesting applications of this subject. The volume should be accessible to a serious reader with some knowledge of matrix theory, who is prepared to make an effort to understand it. The writer has lectured from this volume to mathematically unsophisticated audiences with good results.

The writer thanks R. C. Thompson for his critical reading of the manuscript, which disclosed a number of gaps and inaccuracies. He also thanks Doris Burrell for her painstaking efforts in preparing the typed manuscript.

Chapter I

Representations of Arbitrary Groups

1 Reducibility

Let $\mathscr{A} = \{A\}$ be a set (finite or infinite) of $n \times n$ matrices over the complex field C. Then $\mathscr{A}$ is said to be *reducible* if fixed positive integers, p, q and a fixed nonsingular matrix S exist such that for each $A \in \mathscr{A}$,

$$S^{-1}AS = \begin{pmatrix} A_{11} & 0 \\ A_{21} & A_{22} \end{pmatrix}, \tag{1}$$

where A_{11} is $p \times p$, A_{21} $q \times p$, and A_{22} $q \times q$. Otherwise $\mathscr{A}$ is said to be *irreducible*. If the form (1) can be achieved with $A_{21} = 0$ as well for all $A \in \mathscr{A}$, then $\mathscr{A}$ is said to be *fully reducible*. Thus if T is any $n \times n$ nonsingular matrix, then $T^{-1}\mathscr{A}T = \{T^{-1}AT\}$ is reducible if and only if $\mathscr{A}$ is reducible, and fully reducible if and only if $\mathscr{A}$ is fully reducible.

Examples

(a) The set consisting of a single $n \times n$ matrix A alone, $n > 1$, is reducible. In fact there is a nonsingular matrix S such that $S^{-1}AS$ is lower triangular.

(b) The set $\mathscr{A} = \left\{ \begin{pmatrix} 1 & 1 \\ 0 & 1 \end{pmatrix}, \begin{pmatrix} 1 & 0 \\ 1 & 1 \end{pmatrix} \right\}$ is irreducible.

(c) Let $\mathscr{A}$ be a set of $n \times n$ column stochastic matrices (having all column sums 1). Then $\mathscr{A}$ is reducible. To see this, choose

$$S = \begin{bmatrix} 0 & -1 & -1 & \cdots & -1 \\ 0 & 1 & 0 & \cdots & 0 \\ 0 & 0 & 1 & \cdots & 0 \\ & & \cdots & & \\ 0 & 0 & 0 & \cdots & 1 \end{bmatrix}.$$

Then if $A \in \mathscr{A}$, the first row of $S^{-1}AS$ is the first unit vector.

(d) Let $\mathscr{A} = \{A\}$, where each matrix A has the form $A = \begin{pmatrix} U & V \\ V & U \end{pmatrix}$, with respect to some fixed partitioning. Then $\mathscr{A}$ is reducible. In fact,

$$\begin{pmatrix} I & I \\ 0 & I \end{pmatrix} \begin{pmatrix} U & V \\ V & U \end{pmatrix} \begin{pmatrix} I & -I \\ 0 & I \end{pmatrix} = \begin{pmatrix} U+V & 0 \\ V & U-V \end{pmatrix}.$$

(e) Let $\mathscr{A}$ consist of 2×2 matrices of the form $\begin{pmatrix} x & -y \\ y & x \end{pmatrix}$. Then $\mathscr{A}$ is reducible. In fact,

$$\begin{pmatrix} 1 & i \\ 0 & 1 \end{pmatrix} \begin{pmatrix} x & -y \\ y & x \end{pmatrix} \begin{pmatrix} 1 & -i \\ 0 & 1 \end{pmatrix} = \begin{pmatrix} x+iy & 0 \\ y & x-iy \end{pmatrix}.$$

(f) The set consisting of the matrix $\begin{pmatrix} 1 & 1 \\ 0 & 1 \end{pmatrix}$ alone is reducible, but not fully reducible.

2   Schur's Lemma

The basic result on irreducible sets of matrices is Schur's lemma, which states the following:

Theorem 1 (Schur's lemma). *Let $\mathscr{A} = \{A\}$, $\mathscr{B} = \{B\}$ be irreducible sets of $n \times n$ matrices, $m \times m$ matrices respectively. Let M be fixed $m \times n$ matrix which determines a 1–1 correspondence between $\mathscr{A}$ and $\mathscr{B}$ such that*

$$MA = BM.$$

Then either M = 0, or m = n and M is nonsingular.

Proof. Suppose that the rank of M is r, and write

$$M = P \begin{pmatrix} 0 & 0 \\ I_r & 0 \end{pmatrix} Q,$$

where P, Q are nonsingular and P is $m \times m$, Q $n \times n$. Then for each $A \in \mathscr{A}$ and the corresponding $B \in \mathscr{B}$,

$$\begin{pmatrix} 0 & 0 \\ I_r & 0 \end{pmatrix} QAQ^{-1} = P^{-1}BP \begin{pmatrix} 0 & 0 \\ I_r & 0 \end{pmatrix}. \tag{2}$$

Put

$$QAQ^{-1} = \begin{bmatrix} A_{11} & A_{12} \\ A_{21} & A_{22} \end{bmatrix}, \quad P^{-1}BP = \begin{bmatrix} B_{11} & B_{12} \\ B_{21} & B_{22} \end{bmatrix},$$

where the partitioning is that imposed by (2). It is easily verified that A_{12} is $r \times (n-r)$, and $B_{12}(m-r) \times r$. Then (2) implies that

$$\begin{bmatrix} 0 & 0 \\ A_{11} & A_{12} \end{bmatrix} = \begin{bmatrix} B_{12} & 0 \\ B_{22} & 0 \end{bmatrix}.$$

Hence $A_{12} = 0$, $B_{12} = 0$, an impossibility, since $\mathscr{A}$ and $\mathscr{B}$ are irreducible. Since this can happen only vacuously, we must have that either $r = 0$ (in which case $M = 0$) or $r = m = n$ (in which case M is nonsingular). This completes the proof of the theorem. $\square$

An important corollary is the following:

Corollary 1. *Suppose that* M *commutes with each matrix of the irreducible set* $\mathscr{A}$. *Then* M *is scalar.*

Proof. Let λ be any eigenvalue of M. Then $M - \lambda I$ is singular and also commutes with each matrix of $\mathscr{A}$. Schur's lemma now implies that $M - \lambda I$ must be 0. This completes the proof of the corollary. $\square$

3   Representations

A representation α of degree n of a group G is a homomorphism of G into $GL(n, C)$; that is, into a group of $n \times n$ nonsingular matrices with arbitrary complex entries. The representation is *faithful* if it is an isomorphism. We shall write

$$\alpha = \{A(x) : x \in G\}$$

where the matrices $A(x)$ obey the homomorphism rule

$$A(xy) = A(x)A(y), \quad x, y \in G.$$

In particular, $A(1) = I_n$, $A(x^{-1}) = A(x)^{-1}$, $x \in G$, where 1 is the unit element of G.

Two representations α and β are said to be *equivalent* (written $\alpha \sim \beta$) if there is a fixed $M \in GL(n, C)$ such that

$$\beta = \{M^{-1}A(x)M : x \in G\}.$$

It is readily verified that "$\sim$" is an equivalence relationship. α is said to be reducible, irreducible, etc., if the corresponding set of matrices is reducible, irreducible, etc.

Let $\alpha = \{A(x) : x \in G\}$, $\beta = \{B(x) : x \in G\}$, be representations of G. Then the *sum* of the representations (denoted by $\alpha + \beta$) is the representation

$$\alpha + \beta = \left\{ A(x) + B(x) = \begin{pmatrix} A(x) & 0 \\ 0 & B(x) \end{pmatrix} : x \in G \right\}.$$

Notice that in general $\alpha + \beta \neq \beta + \alpha$, but always $\alpha + \beta \sim \beta + \alpha$.

A representation of degree one is sometimes referred to as a one-dimensional representation, or a linear representation.

The representation which assigns the identity matrix I to each group element is called the trivial representation. Since the kernel of α is a normal subgroup of G, any nontrivial representation of a simple group must be faithful.

There exist groups which have only the trivial representation; for example, the group $G = \{a, b, c, d\}$ generated by the four elements a, b, c, d with defining relations

$$b^{-1}ab = a^2, \quad c^{-1}bc = b^2, \quad d^{-1}cd = c^2, \quad a^{-1}da = d^2,$$

first discussed by Higman. The proofs that G is nontrivial and that all its representations are trivial, are quite difficult. See papers [20] and [12] for proofs.

Examples

(a) Let $G = \{x\}$, the infinite cyclic group generated by x. Let A be any non-singular $n \times n$ matrix. Then the representation obtained by assigning A to x is faithful if and only if A is not of finite period.

(b) Let G be the group of order 4 generated by elements x, y such that $x^2 = y^2 = 1$, $xy = yx$. Then

$$x \to \begin{pmatrix} 1 & 0 \\ 0 & -1 \end{pmatrix}, \quad y \to \begin{pmatrix} -1 & 0 \\ 0 & 1 \end{pmatrix}$$

is a faithful representation of G of degree 2, and is reducible. G does not have a faithful irreducible representation. In fact, a later theorem will imply that an abelian group has a faithful irreducible representation if and only if it is cyclic.

(c) The matrices

$$\begin{pmatrix} 1 & 0 & 0 \\ 0 & 1 & 0 \\ 0 & 0 & 1 \end{pmatrix}, \quad \begin{pmatrix} 1 & 0 & 0 \\ 0 & 0 & 1 \\ 0 & 1 & 0 \end{pmatrix}, \quad \begin{pmatrix} 0 & 1 & 0 \\ 1 & 0 & 0 \\ 0 & 0 & 1 \end{pmatrix},$$

$$\begin{pmatrix} 0 & 1 & 0 \\ 0 & 0 & 1 \\ 1 & 0 & 0 \end{pmatrix}, \quad \begin{pmatrix} 0 & 0 & 1 \\ 1 & 0 & 0 \\ 0 & 1 & 0 \end{pmatrix}, \quad \begin{pmatrix} 0 & 0 & 1 \\ 0 & 1 & 0 \\ 1 & 0 & 0 \end{pmatrix},$$

form a faithful representation of S_3 of degree 3, which is reducible.

(d) The matrices

$$\begin{pmatrix} 1 & 0 \\ 0 & 1 \end{pmatrix}, \quad \begin{pmatrix} 0 & 1 \\ 1 & 0 \end{pmatrix}, \quad \begin{pmatrix} -1 & -1 \\ 0 & 1 \end{pmatrix},$$

$$\begin{pmatrix} 0 & 1 \\ -1 & -1 \end{pmatrix}, \quad \begin{pmatrix} -1 & -1 \\ 1 & 0 \end{pmatrix}, \quad \begin{pmatrix} 1 & 0 \\ -1 & -1 \end{pmatrix},$$

form a faithful irreducible representation of S_3 of degree 2.

(e) Let G be the free group of rank 2 with free generators x, y. Then

$$x \to \begin{pmatrix} 1 & 2 \\ 0 & 1 \end{pmatrix}, \quad y \to \begin{pmatrix} 1 & 0 \\ 2 & 1 \end{pmatrix}$$

is a faithful irreducible representation of G of degree 2. (See [2] or [17] for a proof.)

(f) Let G be the multiplicative group of the nonzero complex numbers. If $z = x + iy$ is any element of G, where x and y are real, then

$$z \to \begin{pmatrix} x & -y \\ y & x \end{pmatrix}$$

is a faithful representation of G of degree 2, which is reducible.

(g) The group of planar rotations is faithfully represented by

$$\left\{ \begin{pmatrix} \cos\theta & -\sin\theta \\ \sin\theta & \cos\theta \end{pmatrix} : \theta \text{ real} \right\}$$

and also by

$$\{(e^{i\theta}) : \theta \text{ real}\}.$$

(h) Let G be the octahedral group, generated by elements x, y with defining relations

$$x^2 = y^3 = (xy)^4 = 1.$$

Then G is of order 24, and

$$x \rightarrow \begin{pmatrix} -1 & 0 \\ 0 & 1 \end{pmatrix}, \quad y \rightarrow \begin{pmatrix} -\frac{1}{2} & -\frac{3}{4} \\ 1 & -\frac{1}{2} \end{pmatrix}$$

is an irreducible representation of G which is *not* faithful. The octahedral group does not have a faithful representation of degree 2. The representation above is in fact a faithful representation of the dihedral group D_3 of order 6.

4 Character of a Representation

The *character* of α is the set of numbers

$$\chi_\alpha = \{\chi_\alpha(x) : x \in G\},$$

where

$$\chi_\alpha(x) = \text{tr}(A(x)) = \sum_{i=1}^{n} a_{ii}(x).$$

Then $\chi_\alpha(x)$ is also the sum of the eigenvalues of $A(x)$. Equivalent representations have the same character, since the trace is a similarity invariant.

Notice that if x is of finite period, then

$$\chi_\alpha(x^{-1}) = \overline{\chi_\alpha(x)}, \quad x \in G.$$

The reason for this is that in this case $A(x)$ is also of finite period, and hence its eigenvalues are all roots of unity. Since the eigenvalues of $A(x^{-l}) = A(x)^{-1}$ are the reciprocals of those of $A(x)$, and since the reciprocal of a number of modulus 1 is its complex conjugate, the result follows.

5 Theorems of Burnside, Frobenius, and Schur

The most important single fact about irreducible representations is undoubtedly the following theorem:

Theorem 2 (Burnside). *Let*

$$\alpha = \{A(x) : x \in G\}$$

be an irreducible representation of degree n *of the group* G. *Then any relationship*

$$\sum_{p,q} \tau_{pq} a_{pq}(x) = 0$$

can hold for all x $\in$ G *if and only if* $\tau_{pq} = 0$ *for all* p, q.

An equivalent formulation of this theorem (which may be regarded as a corollary) is the following:

Corollary 2. *Let α be an irreducible representation of degree* n *of the group* G. *Then the dimension of the vector space spanned by the matrices of α is precisely* n^2.

We shall give the proof by Frobenius and Schur of this theorem, and later on their generalization of it. Before doing so, however, we insert a proof of the equivalence of the two forms of the theorem, which we state as a lemma:

Lemma 1. *Let* W *be a vector space of dimension* n, V *a subspace of* W *of dimension* s. *Then scalars* $c_1, c_2, \ldots, c_n$, *which are not all* 0, *exist such that*

$$\sum_{p=1}^{n} c_p \tau_p = 0$$

for all vectors $\tau = (\tau_1, \tau_2, \ldots, \tau_n)$ *of* V, *if and only if* s < n.

Proof. Let $y_1, y_2, \ldots, y_s$ be a basis for V. Extend this basis by vectors $y_{s+1}, \ldots, y_n$ to a basis for W. Then any vector $\tau \in W$ has a unique representation with respect to this basis as

$$\tau = \sum_{p=1}^{n} \tau_p y_p = (\tau_1, \tau_2, \ldots, \tau_n),$$

and $\tau \in V$ if and only if $\tau_{s+1} = \cdots = \tau_n = 0$. A vector $c = (c_1, c_2, \ldots, c_n)$ is a solution of

$$(c, \tau) = \sum_{p=1}^{n} c_p \tau_p = 0, \quad \text{all } \tau \in V, \tag{3}$$

if and only if

$$(c, y_i) = 0, \quad 1 \leq i \leq s.$$

This implies that $c_i = 0$, $1 \leq i \leq s$. Thus c is a solution of (3) if and only if

$$c = (0, 0, \ldots, 0, c_{s+1}, \ldots, c_n);$$

and hence a nonzero solution of (3) exists if and only if $s < n$. This completes the proof of the lemma, and demonstrates the equivalence of the two forms of Burnside's theorem. $\square$

We now go on to the proof of theorem 2. For notational convenience, we denote the matrices of α by A, B, etc. and their totality by $\mathscr{A}$, so that their dependence on the group elements is not indicated. Suppose that n^2 constants k_{qp}

exist which are not all 0 such that

$$\sum_{p,q} k_{qp} a_{pq} = 0, \quad \text{all } A \in \mathscr{A}.$$

Set $K = (k_{pq})$. Then the above is equivalent to

$$\text{tr}(KA) = 0, \quad \text{all } A \in \mathscr{A}. \tag{4}$$

The totality of matrices K satisfying (4) is a vector space W of dimension $s \leq n^2$. Let $K_1, K_2, \ldots, K_s$ be a basis for W. Then every matrix K satisfying (4) is representable as a linear combination of $K_1, K_2, \ldots, K_s$. Furthermore if $K \in W$ then for each $A \in \mathscr{A}$, $KA \in W$, since

$$\text{tr}(KA \cdot B) = \text{tr}(K \cdot AB) = 0, \quad \text{all } B \in \mathscr{A}.$$

Hence there are s constants $r_1, r_2, \ldots, r_s$ (which depend on A) such that

$$KA = \sum_{p=1}^{s} r_p K_p;$$

and these are uniquely, determined, since the matrices $K_1, K_2, \ldots, K_s$ are linearly independent. In particular we have that

$$K_p A = \sum_{q=1}^{s} r_{pq} K_q, \quad 1 \leq p \leq s. \tag{5}$$

Put

$$K_p = (k_{ij}^p), \quad 1 \leq p \leq s.$$

Then the s equations (5) become

$$\sum_{l=1}^{n} k_{il}^p a_{lj} = \sum_{q=1}^{s} r_{pq} K_{ij}^q, \quad 1 \leq p \leq s, 1 \leq i, j \leq n. \tag{6}$$

Denote the $s \times s$ matrix

$$\begin{bmatrix} r_{11} & r_{12} & \cdots & r_{1s} \\ r_{21} & r_{22} & \cdots & r_{2s} \\ & \cdots & & \\ r_{s1} & r_{s2} & \cdots & r_{ss} \end{bmatrix}$$

by R, and the $s \times n$ matrix

$$\begin{bmatrix} k_{i1}^1 & k_{i2}^1 & \cdots & k_{in}^1 \\ k_{i1}^2 & k_{i2}^2 & \cdots & k_{in}^2 \\ & \cdots & & \\ k_{i1}^s & k_{i2}^s & \cdots & k_{in}^s \end{bmatrix}$$

by P_i, $1 \leq i \leq n$. Then the equations (6) may be written in the form

$$P_i A = RP_i, \quad 1 \leq i \leq n. \tag{7}$$

Now suppose that v_{pq} are s^2 arbitrary numbers such that $V = (v_{pq})$ is nonsingular. If we define

$$K'_p = \sum_{q=1}^{s} v_{pq} K_q, \quad 1 \leq p \leq s$$

then $K'_1, K'_2, \ldots, K'_s$ is also a basis for W. Define R' by

$$R' = VRV^{-1} = (r'_{pq}).$$

Then (5) takes the form

$$K'_p A = \sum_{q=1}^{s} r'_{pq} K'_q, \quad 1 \leq p \leq s.$$

It follows that in our discussion $\mathscr{R} = \{R\}$ may be replaced by any equivalent set, by making the appropriate change of basis in W.

We now appeal to Schur's lemma (theorem 1). Since the matrices $K_1, K_2, \ldots, K_s$ are linearly independent, none of them is 0. Hence not all of the n matrices P_i, $1 \leq i \leq n$ can be 0. It follows that either P_i is a square nonsingular matrix for some i, in which case the sets $\mathscr{A}$ and $\mathscr{R}$ are equivalent, or the set $\mathscr{R}$ is reducible.

Suppose first that P_i is square and nonsingular for some i so that $s = n$ and $\mathscr{A}$ and $\mathscr{R}$ equivalent. Then after a suitable change of basis in W has been performed we can in fact assume that they are the same. Equations (7) now become

$$P_i A = AP_i, \quad 1 \leq i \leq n.$$

Corollary 1 now implies that P_i is scalar:

$$P_i = k_i I, \quad 1 \leq i \leq n.$$

Thus

$$k_{ij}^p = k_i \delta_{pj}.$$

Now from the fact that

$$\sum_{i,j} k_{ji}^p a_{ij} = 0, \quad 1 \leq p \leq n$$

it follows that

$$\sum_{i,j} k_j \delta_{pi} a_{ij} = 0, \quad 1 \le p \le n,$$

$$\sum_{j=1}^{n} k_j a_{pj} = 0, \quad 1 \le p \le n.$$

Hence

$$A \begin{bmatrix} k_1 \\ k_2 \\ \cdots \\ k_n \end{bmatrix} = 0.$$

But A is nonsingular: Thus

$$k_j = 0, \quad 1 \le j \le n$$

which in turn implies that all the matrices P_i are zero, $1 \le i \le n$; a contradiction.

Next suppose that $\mathscr{R}$ is reducible. Then after a suitable change of basis has been performed we may assume that for some positive integer t, every $R \in \mathscr{R}$ has the form

$$R = \begin{pmatrix} S & 0 \\ T & U \end{pmatrix},$$

where the $t \times t$ matrices S are either all 0, or $\{S\}$ is irreducible. If $S = (s_{pq})$, then the first t equations of (5) become

$$K_p A = \sum_{q=1}^{t} s_{pq} K_q, \quad 1 \le p \le t.$$

If $\{S\}$ is irreducible, we conclude as before that the t matrices $K_1, K_2, \ldots, K_t$ are all 0. If the S are all 0, the same conclusion can be drawn directly from the equations

$$K_p A = 0, \quad 1 \le p \le t,$$

since A is nonsingular. In either case we have arrived at a contradiction, since the matrices $K_1, K_2, \ldots, K_t$ are linearly independent and so cannot vanish.

This concludes the proof of the theorem.

The generalization by Frobenius and Schur of Burnside's theorem is as follows:

Theorem 3 (Frobenius, Schur). *Let* $\alpha = \{A(x) : x \in G\}$, $\beta = \{B(x) : x \in G\}$, *... be a set of finitely many irreducible and pairwise inequivalent representations* *of a group* G. *Then any relationship*

$$\sum_{p,q} \tau_{pq} a_{pq}(x) + \sum_{p,q} \eta_{pq} b_{pq}(x) + \cdots = 0$$

can hold for all $x \in G$ *if and only if* $\tau_{pq} = 0$, *all* p, q, $\eta_{pq} = 0$, *all* p, q, *etc.*

Proof. As in the proof of theorem 2, denote the totality of matrices of α by $\mathscr{A} = \{A\}$, the totality of matrices of β by $\mathscr{B} = \{B\}, \ldots$, and assume the existence of a non-trivial relationship

$$\text{tr}(KA) + \text{tr}(LB) + \cdots = 0, \tag{8}$$

where $A, B, \ldots$ correspond to the same group element of G. The totality of solutions of (8) is again a vector space W, consisting of all vectors $(K, L, \ldots)$ satisfying (8). Suppose that W is of dimension s, and let

$$(K_p, L_p, \ldots), \quad 1 \le p \le s$$

be a basis for W. Then corresponding to any solution $(K, L, \ldots)$ of (8) there are constants r_p, $1 \le p \le s$ (which are uniquely determined by the solution) such that

$$(K, L, \ldots) = \sum_{p=1}^{s} r_p (K_p, L_p, \ldots).$$

Hence

$$K = \sum_{p=1}^{s} r_p K_p, \quad L = \sum_{p=1}^{s} r_p L_p, \ldots.$$

As in the proof of theorem 2, if $(K, L, \ldots)$ is a solution of (8) then so is $(KA, LB, \ldots)$ where $A, B, \ldots$ correspond to the same group element. In particular there exist s^2 constants r_{pq} such that

$$K_p A = \sum_{q=1}^{s} r_{pq} K_q, \quad L_p B = \sum_{q=1}^{s} r_{pq} L_q, \ldots, \quad 1 \le p \le s.$$

Let $\mathscr{R}$ denote the totality $\{R\}$, where $R = (r_{pq})$. Then as before $\mathscr{R}$ may be replaced by any equivalent set, and there is no loss of generality in taking $\mathscr{R}$ irreducible.

From the equations

$$K_p A = \sum_{q=1}^{s} r_{pq} K_q, \quad 1 \le p \le s,$$

it follows exactly as before that the matrices $K_1, K_2, \ldots, K_s$ are either all 0, or $\mathscr{R}$ and $\mathscr{A}$ are equivalent. The same also holds for $\mathscr{B}, \ldots$. But the representations

$\alpha, \beta, \ldots$ are pairwise inequivalent. Hence $\mathscr{R}$ can be equivalent to at most one of $\mathscr{A}, \mathscr{B}, \ldots$ which we can suppose to be $\mathscr{A}$. It follows then that

$$L_p = 0, \ldots, \quad 1 \leq p \leq s.$$

Hence

$$\mathrm{tr}(K_p A) = 0, \quad 1 \leq p \leq s, \text{ all } A \in \mathscr{A}.$$

Since α is irreducible, all the K_p must be 0 as well, by Burnside's theorem. Thus we have arrived at a contradiction and the proof is concluded.   □

6 An Application to Matrix Groups

If G is any group, we say that G is of *exponent* r if $x^r = 1$ for all $x \in G$ (equivalently, $G^r = \{1\}$). Among the many important applications of theorem 2, the following is one of the most interesting:

Theorem 4 (Burnside). *Let* G *be a subgroup of* GL(n, C) *of exponent* r. *Then* G *is finite.*

Proof. Suppose first that G is irreducible. If $A \in G$, then $\mathrm{tr}(A)$ is the sum of n rth roots of 1, and so assumes only finitely many values (certainly not more than r^n). Therefore the set

$$\{\mathrm{tr}(A) : A \in G\}$$

contains only finitely many distinct elements.

Since G is irreducible, there are elements $A_1, A_2, \ldots, A_{n^2}$ of G which are linearly independent. Let A be any element of G, and put $A = (a_{ij})$, $A_k = (a_{ij}^k)$, $1 \leq k \leq n^2$. Then

$$\mathrm{tr}(A_k A) = \sum_{i,j} a_{ij}^k a_{ji}, \quad 1 \leq k \leq n^2.$$

Regarding these as n^2 linear equations in the n^2 unknowns a_{ji}, we see that the rows of the coefficient matrix are linearly independent and hence that a unique solution exists, depending only on the coefficient matrix and the values $\mathrm{tr}(A_k A)$, $1 \leq k \leq n^2$. Since there are only finitely many possibilities for these, there are only finitely many possibilities for the a_{ji}. Thus the group G is finite in this case.

We now proceed by induction on n. The theorem is certainly true when $n = 1$. Suppose the theorem proved for all dimensions $< n$. If G is irreducible, the discussion above shows that the theorem is true. If G is reducible, we may assume that G is already in reduced form; that is, that there are positive integers p, q such that every $A \in G$ has the form

$$A = \begin{pmatrix} B & 0 \\ C & D \end{pmatrix},$$

where B is $p \times p$, C is $q \times p$ and D is $q \times q$. Then the groups $\{B\}$ and $\{D\}$ are finite, by the induction hypothesis, since they are certainly of exponent r. Define

$$G_1 = \{A \in G : B = I\}, \quad G_2 = \{A \in G : D = I\}.$$

Then G_1, G_2 are normal subgroups of G of finite index. The same is therefore true of $G_1 \cap G_2$. Let $A \in G_1 \cap G_2$. Then

$$A = \begin{pmatrix} I & 0 \\ C & I \end{pmatrix}.$$

Since

$$A^r = \begin{pmatrix} I & 0 \\ rC & I \end{pmatrix} = I,$$

$C = 0$. Hence $G_1 \cap G_2 = \{I\}$, and it follows that G is finite. This completes the proof. $\qquad\square$

It is worth pointing out that an almost identical discussion shows that if G is any subgroup of $GL(n, C)$ with only a finite number of conjugacy classes k, then G is a finite group. Again the relevant point is that the set $\{\operatorname{tr}(A) : A \in G\}$ contains only finitely many distinct terms.

Notice that theorem 4 is not necessarily true for arbitrary groups. A counterexample is furnished by the infinite abelian group

$$G_r = C_r \times C_r \times C_r \times \cdots ,$$

where C_r is a cyclic group of order r, since G_r is of exponent r. It also follows that G_r has no faithful representation as a finite-dimensional matrix group.

Burnside conjectured that a finitely generated group of exponent r is necessarily finite. Although the general conjecture is now known to be false, it is true for $r = 1, 2, 3, 4, 6$.

Schur has shown that if G is a finitely generated subgroup of $GL(n, C)$ such that each element of G is of finite order, then G is finite; and Selberg has shown that any finitely generated matrix group over C has a subgroup of finite index with no elements of finite order.

Finally, we note that it is known that any torsion-free group (that is, with no elements of finite order) can be embedded in a group with just two conjugacy classes, so that infinite groups exist with only finitely many conjugacy classes.

7   Further Consequences

Most of our applications will be made to finite groups. Before doing so, however, we mention several other consequences of Schur's lemma and the Burnside-Frobenius-Schur theorems.

Suppose that $\alpha = \{A(x) : x \in G\}$, $\beta = \{B(x) : x \in G\}$, are representations of G (not necessarily of the same degree). A matrix P such that

$$PA(x) = B(x)P, \quad x \in G$$

will be said to *link* α and β. Then Schur's lemma easily implies

Corollary 3. *If* P *links two inequivalent irreducible representations then* P $= 0$.

Corollary 4. *If* P *links two irreducible representations of the same degree and* P *is singular then* P $= 0$.

We also have from corollary 1:

Theorem 5. *Every irreducible representation of an abelian group is of degree* 1.

Theorem 3 easily implies

Theorem 6. *The characters of a finite set of pairwise inequivalent irreducible representations of a group G are linearly independent.*

Proof. Suppose that the representations are $\alpha = \{A(x) : x \in G\}$, $\beta = \{B(x) : x \in G\}, \ldots$ with characters $\chi_\alpha, \chi_\beta, \ldots$. Then the relationship

$$\tau_\alpha \chi_\alpha(x) + \tau_\beta \chi_\beta(x) + \cdots = 0, \quad x \in G$$

is equivalent to the relationship

$$\sum_p \tau_\alpha a_{pp}(x) + \sum_p \tau_\beta b_{pp}(x) + \cdots = 0, \quad x \in G,$$

which is possible only if $\tau_\alpha = \tau_\beta = \cdots = 0$, by theorem 3.

Let $\alpha = \{A(x) : x \in G\}$ be a representation of degree n of G. A *reduction* of α is a similarity

$$S^{-1}\alpha S = \{S^{-1}A(x)S : x \in G\}$$

where

$$S^{-1}A(x)S = \begin{pmatrix} A_{11}(x) & 0 & \cdots & 0 \\ A_{21}(x) & A_{22}(x) & \cdots & 0 \\ & & \cdots & \\ A_{p1}(x) & A_{p2}(x) & \cdots & A_{pp}(x) \end{pmatrix}, \quad x \in G,$$

and the representations

$$\alpha_i = \{A_{ii}(x) : x \in G\}, \quad 1 \le i \le p,$$

are irreducible. It is clear that every representation has a reduction. p is the *length* of the reduction, and the α_i are the *irreducible components* of the reduction. □

Then we have the theorem

Theorem 7. *Suppose that the two representations α, β of the group G have the same character. Then any reduction of α has the same length as any reduction of β, and the irreducible components of the two reductions are equivalent, apart from order.*

Proof. Let α_i, $1 \le i \le t$, be a set of pairwise inequivalent irreducible representations such that every irreducible component occurring in the two reductions is equivalent to some α_i, and let χ_i be the character of α_i. Suppose that α_i is equivalent to r_i irreducible components of the reduction of α, and to s_i irreducible components of the reduction of β. Since α and β have the same character, it follows that

$$\sum_{i=1}^{t} r_i \chi_i(x) = \sum_{i=1}^{t} s_i \chi_i(x), \quad x \in G,$$

$$\sum_{i=1}^{t} (r_i - s_i) \chi_i(x) = 0, \quad x \in G.$$

Then theorem 6 implies that

$$r_i = s_i, \quad 1 \le i \le t.$$

This completes the proof of the theorem. □

Some important consequences of theorem 7 are:

Corollary 5. *Two reductions of a representation have the same length and (apart from order) equivalent irreducible components.*

Corollary 6. *Two irreducible representations of a group are equivalent if and only if their characters are the same.*

Chapter II

Representations of Finite Groups

8   Theorems of Maschke and Schur

When G is finite, we denote its order by h and the number of its conjugacy classes by k.

We first determine the irreducible representations of a finite abelian group. We have

Theorem 8. *The irreducible representations of a finite abelian group* G *of order* h *are all of degree 1, and there are exactly* h *of them.*

Proof. We need only prove the last statement. Let $G = G_1 \times G_2 \times \cdots \times G_s$ be a decomposition of G as the direct product of the cyclic groups G_i of prime power order $p_i^{e_i}$. Then every representation of G of degree 1 is obtained by assigning the value ζ_i to the generator g_i of G_i, where ζ_i is any $p_i^{e_i}$ th root of 1; and two such representations are the same if and only if they each have the same ζ_i for each i. Hence the total number of representations is just

$$\prod_{i=1}^{s} p_i^{e_i} = h. \qquad \square$$

The next theorem indicates that the study of the representations of a finite group is particularly simple.

Theorem 9 (**Maschke**). *A reducible representation of a finite group* G *is fully reducible.*

Proof. Let

$$\alpha = \left\{ M(x) = \begin{bmatrix} A(x) & 0 \\ C(x) & D(x) \end{bmatrix} : x \in G \right\}$$

be reduced and equivalent to the given reducible representation. Then $M(xy) = M(x)M(y)$ for all $x, y \in G$, which implies that

$$A(xy) = A(x)A(y), \quad D(xy) = D(x)D(y), \quad C(xy) = C(x)A(y) + D(x)C(y).$$

We write the last relationship as

$$C(x) = C(xy)A(y)^{-1} - D(x)C(y)A(y)^{-1}$$
$$= C(xy)A(xy)^{-1}A(x) - D(x)C(y)A(y)^{-1}.$$

Put

$$C = \sum_{y \in G} C(y)A(y)^{-1}.$$

Then summing over all $y \in G$, we find that for all $x \in G$,

$$hC(x) = CA(x) - D(x)C.$$

Put

$$T = \begin{bmatrix} I & 0 \\ \dfrac{1}{h}C & I \end{bmatrix}.$$

Then it is readily verified that

$$T^{-1}M(x)T = \begin{bmatrix} A(x) & 0 \\ 0 & D(x) \end{bmatrix}.$$

Hence $T^{-1}\alpha T$ is fully reduced, and the theorem follows. $\square$

Another result of interest of the same kind is the following theorem:

Theorem 10 (**Schur**). *Every representation of a finite group has an equivalent unitary representation (one by unitary matrices).*

Proof. Suppose that $\alpha = \{A(x) : x \in G\}$ is the given representation, and put

$$A = \sum_{y \in G} A(y)A(y)^*.$$

Then A is positive definite, and $A(x)AA(x)^* = A$, for each $x \in G$. Write $A = U^*DU$, where U is unitary and D diagonal with positive diagonal entries. Then $D^{1/2}$ (the positive square root of D), is well-defined. Set $T = D^{-1/2}U$. We have for each $x \in G$ that

$$(TA(x)T^{-1})(TA(x)T^{-1})^* = (D^{-1/2}UA(x)U^*D^{1/2})(D^{1/2}UA(x)^*U^*D^{-1/2})$$
$$= D^{-1/2}UA(x)U^*DUA(x)^*U^*D^{-1/2}$$
$$= D^{-1/2}UA(x)AA(x)^*U^*D^{-1/2}$$
$$= D^{-1/2}UAU^*D^{-1/2}$$
$$= D^{-1/2}DD^{-1/2} = I.$$

Thus each matrix of the representation $T\alpha T^{-1}$ is unitary, and the result follows. $\square$

We note one or two consequences of theorems 8 and 9:

Corollary 7. *Every representation of a finite abelian group has an equivalent diagonal representation.*

Corollary 8. *A matrix of finite period is similar to a diagonal matrix.*

9   Characters of Finite Groups

We go on now to develop some formulas for the characters of finite groups.

Let $\alpha = \{A(x) : x \in G\}$, $\beta = \{B(x) : x \in G\}$ be two irreducible representations of the finite group G, and suppose that α is of degree m, β of degree n. Let U be an arbitrary $m \times n$ matrix, and form

$$V = \sum_{x \in G} A(x)UB(x^{-1}).$$

Let y be any element of G. Then

$$A(y)VB(y)^{-1} = \sum_{x \in G} A(yx)UB((yx)^{-1}) = V,$$

since yx runs over G as x does. It follows that V links α and β. Comparing elements in the (i, j) position, we find that

$$v_{ij} = \sum_{x \in G} \sum_{p,q} a_{ip}(x)u_{pq}b_{qj}(x^{-1}).$$

First consider the case when α and β are inequivalent. Then corollary 3 implies that $V = 0$, and since U was arbitrary, we find that

$$\sum_{x \in G} a_{ip}(x) b_{qj}(x^{-1}) = 0, \quad \alpha \not\sim \beta. \tag{9}$$

In particular, setting $p = i$, $q = j$ and summing over i, j we obtain

$$\sum_{x \in G} \chi_\alpha(x) \chi_\beta(x^{-1}) = 0, \quad \alpha \not\sim \beta. \tag{10}$$

Now consider the case when $\alpha = \beta$. Then V commutes with every element of α, and corollary 1 implies that V is scalar; $V = \lambda I$. To determine λ we compute the trace:

$$\mathrm{tr}(V) = \mathrm{tr}\left(\sum_{x \in G} A(x) U A(x^{-1})\right) = \sum_{x \in G} \mathrm{tr}(A(x) U A(x)^{-1}) = h\,\mathrm{tr}(U).$$

Hence

$$\lambda = \frac{h}{m}\,\mathrm{tr}(U),$$

and so

$$\frac{h}{m}(u_{11} + u_{22} + \cdots + u_{mm})\delta_{ij} = \sum_{x \in G}\sum_{p,q} a_{ip}(x) u_{pq} a_{qj}(x^{-1}).$$

Comparing coefficients, we find that

$$\sum_{x \in G} a_{ip}(x) a_{qj}(x^{-1}) = \frac{h}{m}\delta_{ij}\delta_{pq}. \tag{11}$$

Setting $p = i$, $q = j$ and summing as before over all i, j we find that

$$\sum_{x \in G} \chi_\alpha(x) \chi_\alpha(x^{-1}) = h. \tag{12}$$

Formulas (10) and (12) are known as the orthogonality relationships of the first kind.

Formula (12) may be generalized slightly. Let y be any element of G. If we multiply (11) by $a_{pl}(y)$ and sum over all p we obtain

$$\sum_{x \in G} a_{qj}(x^{-1}) \left\{ \sum_{p=1}^{m} a_{ip}(x) a_{pl}(y) \right\} = \frac{h}{m} \delta_{ij} \sum_{p=1}^{m} \delta_{pq} a_{pl}(y),$$

$$\sum_{x \in G} a_{qj}(x^{-1}) a_{il}(xy) = \frac{h}{m} \delta_{ij} a_{ql}(y). \tag{13}$$

Choosing $q = j, l = i$ and summing over all i, j, we obtain

$$\sum_{x \in G} \chi_{\ddot{a}}(xy) \chi_{\alpha}(x^{-1}) = \frac{h}{m} \chi_{\alpha}(y). \tag{14}$$

Equivalent forms of (14) are

$$\sum_{x \in G} \chi_{\alpha}(yx) \chi_{\alpha}(x^{-1}) = \frac{h}{m} \chi_{\alpha}(y),$$

$$\sum_{x \in G} \chi_{\alpha}(yx^{-1}) \chi_{\alpha}(x) = \frac{h}{m} \chi_{\alpha}(y). \tag{15}$$

10   A Divisibility Theorem

We shall use (15) to derive a significant result connecting the degree of α with the order G, which we state as a theorem:

Theorem 11. *The degree of an irreducible representation of a finite group divides the order of the group.*

Proof. Let us denote the elements of G by $x_1, x_2, \ldots, x_h$. Set

$$\tau_q = \chi_{\alpha}(x_q), \quad c_{pq} = \chi_{\alpha}(x_p x_q^{-1}), \quad 1 \le p, q \le h.$$

Then (15) becomes

$$\sum_{q=1}^{h} c_{pq} \tau_q = \frac{h}{m} \tau_p, \quad 1 \le p \le h.$$

The τ_q's are not all 0 (since $\chi_{\alpha}(I) = m$, for example) and so this system has a nontrivial solution. It follows that $\frac{h}{m}$ is an eigenvalue of the matrix $C = (c_{pq})$. Now the numbers c_{pq} are algebraic integers, since they are sums of roots of unity (see appendix B). Hence the eigenvalues of C are also algebraic integers (theorem 12 of appendix A) and so $\frac{h}{m}$ is an algebraic integer. But $\frac{h}{m}$ is rational. Hence it is a rational integer, and the proof of the theorem is concluded.   □

Chapter III

Kinds of Representations

11   Classification of Representations

It is interesting and important to know when a representation has an equivalent real
representation; and if it does not, when it is equivalent to its complex conjugate.
With this in mind, we say that a given representation α of degree m is of the *first
kind* if it has an equivalent real representation; of the *second kind*, if it does not
have an equivalent real representation but is equivalent to its complex conjugate $\overline{\alpha}$;
and of the *third kind*, if it is not equivalent to $\overline{\alpha}$.

It is possible to determine the nature of α solely from its character. In order
for α to be of the first or second kinds its character must be real, but we can give
exact criteria.

The material that follows is due to Frobenius and Schur.

We first prove some lemmas.

Lemma 2. *Suppose that the representations α, β are irreducible, unitary, and
equivalent. Then there is a unitary matrix U such that $\beta = U^{-1}\alpha U$.*

Proof. Let S be a matrix such that $\beta = S^{-1}\alpha S$, and suppose that $\alpha =
\{A(x) : x \in G\}$, $\beta = \{B(x) : x \in G\}$. Then for each $x \in G$, $B(x) = S^{-1}A(x)S$,
where the matrices $A(x)$, $B(x)$ are unitary. Taking conjugate transposes and then
inverting, we get $B(x) = S^*A(x)S^{*-1}$, which implies that SS^* commutes with
$A(x)$ for each $x \in G$. Since α is irreducible this implies that $SS^* = cI$, where c is

necessarily positive. Put $U = \frac{1}{\sqrt{c}}S$. Then U is unitary, and $\beta = U^{-1}\alpha U$. This completes the proof of the lemma. $\qquad\qquad\qquad\qquad\qquad\qquad\qquad\qquad\qquad\qquad\quad\square$

Lemma 3. *Let* U *be a symmetric unitary matrix,* p *an arbitrary nonzero integer. Then a symmetric unitary matrix* W *exists such that* $U = W^p$.

Proof. Since U is normal, we may write $U = V^{-1}DV$, where V and D are unitary and D is diagonal. Furthermore we may take

$$D = \zeta_1 I \dotplus \zeta_2 I \dotplus \cdots \dotplus \zeta_t I,$$

where $\zeta_i = \zeta_j$ if and only if $i = j$. The symmetry of U implies that $VV^T D = DVV^T$, which in turn implies that

$$VV^T = V_1 \dotplus V_2 \dotplus \cdots \dotplus V_t,$$

where V_i is unitary. Define

$$D_p = \zeta_1^{\frac{1}{p}} I \dotplus \zeta_2^{\frac{1}{p}} I \dotplus \cdots \dotplus \zeta_t^{\frac{1}{p}} I,$$
$$W = V^{-1} D_p V.$$

It is easily checked that W is symmetric, unitary, and satisfies $U = W^p$. This completes the proof. $\qquad\qquad\qquad\qquad\qquad\qquad\qquad\qquad\qquad\qquad\qquad\quad\square$

It is also quite easy to show that U is symmetric and unitary if and only if it is of the form $V^{-1}DV$, where V is *orthogonal* and D diagonal and unitary.

We now prove

Theorem 12. *Suppose that α is irreducible, and that $\bar{\alpha} = U^{-1}\alpha U$, where* U *is unitary. Then* $U^T = \pm U$. *Furthermore* $U^T = U$ *if and only if α is of the first kind, and* $U^T = -U$ *if and only if α is of the second kind.*

Proof. Conjugating the relationship $\bar{\alpha} = U^{-1}\alpha U$, we obtain $\alpha = \overline{U}^{-1}\bar{\alpha}\overline{U} = U^T \bar{\alpha} U^{T-1}$. Thus

$$\alpha = U^T (U^{-1}\alpha U)U^{T-1},$$
$$U^T U^{-1}\alpha = \alpha U^T U^{-1}.$$

Hence $U^T U^{-1}$ commutes with each matrix of α, and so corollary 1 implies that $U^T U^{-1}$ is scalar, so that $U^T = cU$. Transposing, we also have $U = cU^T$, which implies that $U^T = c^2 U^T$, $c^2 = 1$, $c = \pm 1$. Hence $U^T = \pm U$.

Since α is of the first or second kind by assumption, we need only show that $U^T = U$ if and only if α is of the first kind, to complete the proof.

Suppose that α is of the first kind. Then a matrix S exists such that $S^{-1}\alpha S$ is real. It follows that

$$S^{-1}\alpha S = \overline{S}^{-1}\overline{\alpha}\overline{S} = \overline{S}^{-1}(U^{-1}\alpha U)\overline{S},$$
$$U\overline{S}S^{-1}\alpha = \alpha U\overline{S}S^{-1},$$

Once again corollary 1 implies that $U\overline{S}S^{-1}$ is scalar, so that $U = a S\overline{S}^{-1}$. Consequently

$$U^T = \overline{U}^{-1} = \frac{1}{\overline{a}}S\overline{S}^{-1} = \frac{1}{|a|^2}U.$$

Then $U^T = -U$ implies that $|a|^2 = -1$, an impossibility. It follows that $U^T = U$.

Now suppose that $U^T = U$. By lemma 3, a symmetric unitary matrix W exists such that $U = W^2$. Then the relationship $\overline{\alpha} = U^{-1}\alpha U$ becomes $W\overline{\alpha}W^{-1} = W^{-1}\alpha W$. Hence

$$\overline{W^{-1}\alpha W} = \overline{W\overline{\alpha}W^{-1}} = \overline{W}\alpha\overline{W}^{-1} = W^{-1}\alpha W.$$

Thus $W^{-1}\alpha W$ is real, and the proof of the theorem is concluded.   □

The next theorem supplies an effective criterion for classifying a given representation of a finite group.

Theorem 13. *Suppose that α is an irreducible representation of a finite group* G. *Put*

$$\tau = \frac{1}{h}\sum_{x \in G}\chi_\alpha(x^2).$$

Then $\tau = 1$ if and only if α is of the first kind; $\tau = -1$ if and only if α is of the second kind; and $\tau = 0$ if and only if α is of the third kind.

Proof. By theorem 10, we may assume that α is unitary. Thus if $\alpha = \{A(x) : x \in G\}$ then $A(x)^* = A(x)^{-1}$.

Suppose first that α is of the third kind. Then α and $\overline{\alpha}$ are inequivalent. Choose $\beta = \overline{\alpha}$ in formula (9) to obtain

$$\sum_{x \in G} a_{ip}(x)\overline{a}_{qj}(x^{-1}) = 0,$$

$$\sum_{x \in G} a_{ip}(x)a_{jq}(x) = 0.$$

Choose $p = j, q = i$ and sum over all i, j to obtain

$$\sum_{x \in G}\chi_\alpha(x^2) = 0.$$

The theorem is proved in this case.

Now suppose that α is of the first or second kind, so that α and $\bar{\alpha}$ are equivalent. By lemma 2 we may assume that $\bar{\alpha} = U^{-1}\alpha U$, where U is unitary; and by the preceding theorem, $U^T = U$ if and only if α is of the first kind, while $U^T = -U$ if and only if α is of the second kind. Set $U = (u_{ij})$. Then the relationship $\bar{\alpha} = U^{-1}\alpha U$ implies that

$$\bar{a}_{jq}(x) = \sum_{r,s} u_{rj} a_{rs}(x) u_{sq}, \quad x \in G.$$

Since α is unitary, formula (11) becomes

$$\sum_{x \in G} a_{ip}(x)\bar{a}_{jq}(x) = \frac{h}{m}\delta_{ij}\delta_{pq},$$

and substituting, we obtain

$$\sum_{x \in G}\sum_{r,s} a_{ip}(x)a_{rs}(x)\bar{u}_{rj}u_{sq} = \frac{h}{m}\delta_{ij}\delta_{pq}.$$

First multiply by u_{tj} and sum over j; then multiply by $\bar{u}_{lq}$ and sum over q. The result is

$$\sum_{x \in G} a_{ip}(x)a_{tl}(x) = \frac{h}{m}u_{ti}\bar{u}_{lp}.$$

Now set $p = j, t = j, l = i$ and sum over all i, j to obtain

$$\sum_{x \in G} \chi_\alpha(x^2) = \frac{h}{m}\sum_{i,j} u_{ji}\bar{u}_{ij}.$$

But

$$\sum_{i,j} u_{ji}\bar{u}_{ij} = \begin{cases} m & U^T = U \\ -m & U^T = -U \end{cases}.$$

Hence the theorem is proved for α of the first or second kind, and the proof is concluded. □

An interesting consequence of theorem 12 is the following:

Corollary 9. *Suppose that α is irreducible and of the second kind. Then the degree of α is even.*

Proof. By theorem 12 there is a unitary matrix U such that $U^T = -U$. Hence $(-1)^m \det U = \det U^T = \det U$, $(-1)^m = 1$. Hence m is even, and the proof of the corollary is complete. □

Examples

(a) The matrix group of order 8, whose elements are

$$\pm I, \quad \pm \begin{pmatrix} 0 & 1 \\ -1 & 0 \end{pmatrix}, \quad \pm \begin{pmatrix} i & 0 \\ 0 & -i \end{pmatrix}, \quad \pm \begin{pmatrix} 0 & i \\ i & 0 \end{pmatrix},$$

is irreducible, and the squares of its elements are I (2 times), $-I$ (6 times). Hence $\tau = \frac{1}{8}(2.2 + 6(-2)) = -1$ for this group, so that it is of the second kind.

(b) If the character of a representation α is not real, then α is necessarily of the third kind. In addition, a later result shows that α is of the third kind if and only if its character is not real.

(c) If α is any complex representation of a group G, then $\alpha + \bar{\alpha}$ is of the first kind. For if $\alpha = \{A(x) + iB(x) : x \in G\}$, then for all $x \in G$,

$$\begin{pmatrix} I & iI \\ I & -iI \end{pmatrix} \begin{pmatrix} A(x) & -B(x) \\ B(x) & A(x) \end{pmatrix} \begin{pmatrix} I & iI \\ I & -iI \end{pmatrix}^{-1} = \begin{pmatrix} A(x) + iB(x) & 0 \\ 0 & A(x) - iB(x) \end{pmatrix}.$$

(d) The matrix group consisting of the 2×2 unitary matrices of determinant 1 (that is, all complex matrices of the form

$$\begin{pmatrix} u & -v \\ \bar{v} & \bar{u} \end{pmatrix}, \quad u\bar{u} + v\bar{v} = 1)$$

is irreducible and of the second kind. The irreducibility is readily verified, and

$$\begin{pmatrix} 0 & 1 \\ -1 & 0 \end{pmatrix} \begin{pmatrix} u & -v \\ \bar{v} & \bar{u} \end{pmatrix} \begin{pmatrix} 0 & -1 \\ 1 & 0 \end{pmatrix} = \begin{pmatrix} \bar{u} & -\bar{v} \\ v & u \end{pmatrix} = \overline{\begin{pmatrix} u & -v \\ \bar{v} & \bar{u} \end{pmatrix}}.$$

Since $\begin{pmatrix} 0 & 1 \\ -1 & 0 \end{pmatrix}$ is unitary and skew-symmetric, the group is of the second kind by theorem 12.

Chapter IV

The Principal Results

12 Further Formulas for Characters

We require another character formula, of a different nature, which necessitates some further notation. Let u be a fixed element of the finite group G. Then $[u]$ will denote the conjugacy class determined by u. Also, if u, v are elements of G, we set

$$e(u, v) = \begin{cases} 1 & [u] = [v] \\ 0 & \text{otherwise} \end{cases}$$

$$e(u) = \begin{cases} 1 & u = \text{identity of } G \\ 0 & \text{otherwise.} \end{cases}$$

Let $\alpha = \{A(x) : x \in G\}$ be an irreducible representation of G of degree m, and set

$$C(u) = \sum_{x \in [u]} A(x).$$

Then the matrices $A(x)$, $x \in [u]$, all have the same trace $\chi_\alpha(u)$.

For any $y \in G$, we have

$$
\begin{aligned}
A(y)C(u)A(y)^{-1} &= \sum_{x \in [u]} A(y)A(x)A(y)^{-1} \\
&= \sum_{x \in [u]} A(yxy^{-1}) = C(u),
\end{aligned}
$$

since yxy^{-1} runs over $[u]$ as x does. Hence $C(u)$ commutes with all the matrices of α, and so must have the form λI, by corollary 1. We determine λ by calculating the trace. We have

$$
\lambda m = \operatorname{tr}(C(u)) = h(u)\chi_\alpha(u),
$$

where $h(u)$ is the number of elements in $[u]$. Thus we have

$$
C(u) = \sum_{x \in [u]} A(x) = \frac{h(u)}{m}\chi_\alpha(u) \cdot I. \tag{16}
$$

From (16), we get for the elements in position (p, q),

$$
\sum_{x \in [u]} a_{pq}(x) = \frac{h(u)}{m}\chi_\alpha(u)\delta_{pq}. \tag{17}
$$

13   The Matrix Associated With a Representation

We now associate a matrix with the irreducible representation α of degree m. Let us denote the elements of G by $x_1, x_2, \ldots, x_h$, and define

$$
M(\alpha) = (m/h)^{1/2}
\begin{bmatrix}
a_{11}(x_1) & a_{11}(x_2) & \cdots & a_{11}(x_h) \\
a_{12}(x_1) & a_{12}(x_2) & \cdots & a_{12}(x_h) \\
& & \cdots & \\
a_{mm}(x_1) & a_{mm}(x_2) & \cdots & a_{mm}(x_h)
\end{bmatrix}.
$$

Thus $M(\alpha)$ is an $m^2 \times h$ matrix.

Let us also define the (contragredient) representation α^* as follows:

$$
\alpha^* = \{A(x^{-1})^T : x \in G\}.
$$

Then α and α^* are of the same degree m, and α^* is also irreducible. $M(\alpha^*)$ is then defined accordingly.

The row of $M(\alpha)$ in position (p, q) is

$$R_{p,q} = (m/h)^{1/2}(a_{pq}(x_1)a_{pq}(x_2)\cdots a_{pq}(x_h))$$

and the row of $M(\alpha^*)$ in position (i, j) is

$$S_{i,j} = (m/h)^{1/2}(a_{ji}(x_1^{-1})a_{ji}(x_2^{-1})\cdots a_{ji}(x_h^{-1})).$$

The inner product of these rows is then

$$R_{p,q} \cdot S_{i,j} = \frac{m}{h} \sum_{x \in G} a_{pq}(x)a_{ji}(x^{-1})$$

$$= \delta_{ip}\delta_{jq},$$

by (11). This easily implies that

$$M(\alpha)M(\alpha^*)^T = I_{m^2}. \tag{18}$$

Similarly, if β is an irreducible representation of G which is not equivalent to α, then (9) implies that

$$M(\alpha)M(\beta^*)^T = 0, \quad \alpha \not\sim \beta. \tag{19}$$

Now suppose given a number of pairwise inequivalent irreducible representations $\alpha_1, \alpha_2, \ldots, \alpha_l$ of G, where α_i is of degree m_i; and set

$$M = \begin{bmatrix} M(\alpha_1) \\ M(\alpha_2) \\ \cdots \\ M(\alpha_l) \end{bmatrix}, \quad M_1 = \begin{bmatrix} M(\alpha_1^*) \\ M(\alpha_2^*) \\ \cdots \\ M(\alpha_l^*) \end{bmatrix}.$$

Then (18) and (19) imply that

$$MM_1^T = I_s,$$

where

$$s = m_1^2 + m_2^2 + \cdots + m_l^2.$$

It follows that the rank of $M \geq s$; and since M has exactly s rows, that

$$\text{rank}(M) = s. \tag{20}$$

This implies that the rows are linearly independent, which is another proof of the Schur-Frobenius theorem when the group is finite.

It follows from (20) that the column rank of M is also s, and since there are h columns,

$$m_1^2 + m_2^2 + \cdots + m_l^2 \leq h. \tag{21}$$

Thus if $\alpha_1, \alpha_2, \ldots, \alpha_l$ are any l pairwise inequivalent irreducible representations of the finite group G, then $l \leq h$. We have proved therefore

Theorem 14. *There are at most finitely many pairwise inequivalent irreducible representations of a given finite group* G.

Let us define the *rank* of a representation α as the dimension of the vector space spanned by the matrices of α, or equivalently, as the maximum number of linearly independent matrices of α. Then equivalent representations have the same rank. We shall prove

Lemma 4. *The rank ρ of a representation is equal to the sum of the squares of the degrees of the inequivalent irreducible components of the representation.*

Proof. We can assume the representation fully reduced, and furthermore with equivalent irreducible components identical. No irreducible component need enter with multiplicity more than 1. Thus we need only consider the inequivalent irreducible components, say $\alpha_1, \alpha_2, \ldots, a_l$. The desired rank ρ is clearly just the rank of

$$\begin{bmatrix} M(\alpha_1) \\ M(\alpha_2) \\ \ldots \\ M(\alpha_l) \end{bmatrix},$$

and this (by the previous result) is $m_1^2 + m_2^2 + \cdots + m_l^2$, where m_i is the degree of α_i. This completes the proof of the lemma. $\qquad\qquad\square$

14 The Regular Representation

Further progress depends on the introduction of the regular representation. Define

$$c_{ij}(x_p) = e(x_i x_p x_j^{-1}), \quad 1 \leq p \leq h,$$
$$C(x_p) = (c_{ij}(x_p)).$$

Then

$$\gamma = \{C(x_p) : 1 \leq p \leq h\}$$

forms a representation of degree h of the finite group G by permutation matrices, known as the *regular representation*. (We do not distinguish here between the left and right regular representations.) The verification that γ is a representation is quite simple and is left to the reader.

We shall prove

Lemma 5. *The rank of the regular representation γ is just* h.

Proof. We must show that the matrices of γ are linearly independent. Suppose that τ_p, $1 \le p \le h$, are numbers such that

$$\sum_{p=1}^{h} \tau_p C(x_p) = 0.$$

Then

$$\sum_{p=1}^{h} \tau_p e(x_i x_p x_j^{-1}) = 0.$$

Choose $x_i = 1$. Then

$$0 = \sum_{p=1}^{h} \tau_p e(x_p x_j^{-1}) = \tau_j,$$

and so all the τ_j are 0, $1 \le j \le h$. This completes the proof of the lemma.   □

15   The Principal Formulas

Lemmas 4 and 5 tell us that we have found a set of l pairwise inequivalent irreducible representations of the finite group G of degrees $m_1, m_2, \ldots, m_l$ such that

$$h = m_1^2 + m_2^2 + \cdots + m_l^2.$$

But now (21) tells us that there can be no others which can be added to the set and retain its properties. Hence we have proved

Theorem 15. *Let $\alpha_1, \alpha_2, \ldots, \alpha_l$ be all of the pairwise inequivalent irreducible representations of the finite group* G *of order* h, *and let α_i be of degree* m_i, $1 \le i \le 1$. *Then*

$$h = m_1^2 + m_2^2 + \cdots + m_l^2. \tag{22}$$

We have not yet determined l as a function of the group G. Our next task will be to show that $l = k$, the number of conjugacy classes of G. For this purpose we return to the matrix M. By (22), M must be a square matrix, and hence M_1^T is the inverse of M. Transposing the relationship $MM_1^T = I$, we have $M_1 M^T = I$, and hence $M^T M_1 = I$, since M_1 and M^T are inverses of each other. The pth row of M^T consists of the l vectors

$$(m_r/h)^{1/2}(a_{11}^r(x_p)a_{12}^r(x_p)\cdots a_{m_r m_r}^r(x_p)) \quad 1 \le r \le l$$

where we write

$$\alpha_r = \{A^r(x_p) = (a_{ij}^r(x_p)) : 1 \le p \le h\}.$$

The q^{tn} column of M_1 consists of the l vectors

$$(m_r/h)^{1/2}(a_{11}^r(x_q^{-1})a_{21}^r(x_q^{-1})\cdots a_{m_r m_r}^r(x_q^{-1})), \quad 1 \le r \le l.$$

Forming the inner product of these rows, we get

$$\sum_{r=1}^{l} \sum_{1 \le i,j \le m_r} \frac{m_r}{h} a_{ij}^r(x_p)a_{ji}^r(x_q^{-1}) = \delta_{pq}$$

This may also be written as

$$\sum_{r=1}^{l} \sum_{1 \le i,j \le m_r} m_r a_{ij}^r(x)a_{ji}^r(y^{-1}) = he(xy^{-1}), \tag{23}$$

where x, y are arbitrary elements of G.

Let z be any element of G, and sum (23) over all $x \in [z]$, making use of (17). We obtain

$$\sum_{r=1}^{l} \sum_{1 \le i,j \le m_r} m_r a_{ji}^r(y^{-1}) \left\{ \sum_{x \in [z]} a_{ij}^r(x) \right\} = \sum_{x \in [z]} he(xy^{-1}),$$

$$\sum_{r=1}^{l} \sum_{1 \le i,j \le m_r} a_{ji}^r(y^{-1})h(z)\chi_{\alpha_r}(z)\delta_{ij} = he(z,y),$$

$$\sum_{r=1}^{l} \chi_{\alpha_r}(z)\chi_{\alpha_r}(y^{-1}) = \frac{h}{h(z)}e(z,y), \tag{24}$$

which is known as the orthogonality relationship of the second kind.

Now choose $y = z$ in (24) and sum over all $y \in G$. We obtain

$$\sum_{r=1}^{l} \sum_{y \in G} \chi_{\alpha_r}(y)\chi_{\alpha_r}(y^{-1}) = h \sum_{y \in G} \frac{1}{h(y)}.$$

Let $[y_i]$, $1 \le i \le k$, be the k conjugacy classes of G. Then

$$\sum_{y \in G} \frac{1}{h(y)} = \sum_{i=1}^{k} \sum_{y \in [y_i]} \frac{1}{h(y_i)} = k.$$

Making use of (12), we obtain

$$hl = hk,$$
$$l = k.$$

Hence we have proved

Theorem 16. *There are precisely* k *pairwise inequivalent irreducible representations of the finite group* G, *where* k *is the number of conjugacy classes of* G.

16   Further Results

We now go on to prove some further theorems on representations of finite groups. The theorem that follows is of interest in that it provides an effective criterion for deciding when a given representation of a finite group is irreducible.

Theorem 17. *The representation* α *of the finite group* G *is irreducible if and only if*

$$\sum_{x \in G} |\chi_\alpha(x)|^2 = h.$$

Proof. Let $\alpha_1, \alpha_2, \ldots, \alpha_k$ be the pairwise inequivalent irreducible representations of G, and suppose that α_i occurs with multiplicity v_i in α, $v_i \geq 0$. Then

$$\alpha \sim v_1\alpha_1 + v_2\alpha_2 + \cdots + v_k\alpha_k,$$
$$\chi_\alpha = v_1\chi_{\alpha_1} + v_2\chi_{\alpha_2} + \cdots + v_k\chi_{\alpha_k}.$$

We have

$$\sum_{x \in G} |\chi_\alpha(x)|^2 = \sum_{x \in G} \chi_\alpha(x)\chi_\alpha(x^{-1}) = \sum_{x \in G} \sum_{1 \leq i,j \leq k} v_i v_j \chi_{\alpha_i}(x)\chi_{\alpha_j}(x^{-1})$$
$$= h \sum_{i=1}^{k} v_i^2,$$

by (10) and (12). That is, we have the relationship

$$\sum_{x \in G} |\chi_\alpha(x)|^2 = h \sum_{i=1}^{k} v_i^2 \tag{25}$$

from which the conclusion follows. $\square$

More generally if

$$\alpha \sim \nu_1\alpha_1 + \nu_2\alpha_2 + \cdots + \nu_k\alpha_k,$$
$$\beta \sim \mu_1\alpha_1 + \mu_2\alpha_2 + \cdots + \mu_k\alpha_k,$$

then

$$\sum_{x \in G} \chi_\alpha(x)\overline{\chi_\beta(x)} = h \sum_{i=1}^{k} \nu_i\mu_i.$$

This relationship implies for example that α and β have no irreducible components in common if and only if

$$\sum_{x \in G} \chi_\alpha(x)\overline{\chi_\beta(x)} = 0.$$

Another formula worth noting is that the coefficient ν_i is given by

$$\nu_i = \frac{1}{h} \sum_{x \in G} \chi_\alpha(x)\overline{\chi_{\alpha_i}(x)}.$$

Examples

(a) The representation of S_3 as the multiplicative group of matrices

$$\begin{pmatrix} 1 & 0 & 0 \\ 0 & 1 & 0 \\ 0 & 0 & 1 \end{pmatrix}, \quad \begin{pmatrix} 1 & 0 & 0 \\ 0 & 0 & 1 \\ 0 & 1 & 0 \end{pmatrix}, \quad \begin{pmatrix} 0 & 1 & 0 \\ 1 & 0 & 0 \\ 0 & 0 & 1 \end{pmatrix},$$

$$\begin{pmatrix} 0 & 1 & 0 \\ 0 & 0 & 1 \\ 1 & 0 & 0 \end{pmatrix}, \quad \begin{pmatrix} 0 & 0 & 1 \\ 1 & 0 & 0 \\ 0 & 1 & 0 \end{pmatrix}, \quad \begin{pmatrix} 0 & 0 & 1 \\ 0 & 1 & 0 \\ 1 & 0 & 0 \end{pmatrix}$$

must contain 2 irreducible components, since

$$\frac{1}{h} \sum_{x \in G} |\chi_\alpha(x)|^2 = 2,$$

and the only representation of 2 as the sum of two squares is $2 = 1^2 + 1^2$.

(b) Let G be the octahedral group of order 24, so that G is the group generated by elements x, y with defining relations

$$x^2 = y^3 = (xy)^4 = 1.$$

Then the representation of G given by

$$x \to \begin{pmatrix} 1 & 0 & 0 \\ 0 & 0 & -1 \\ 0 & -1 & 0 \end{pmatrix}, \quad y \to \begin{pmatrix} 0 & 0 & 1 \\ 1 & 0 & 0 \\ 0 & 1 & 0 \end{pmatrix}$$

is irreducible, since it is readily verified that

$$\frac{1}{h} \sum_{x \in G} |\chi_\alpha(x)|^2 = 1.$$

We can obtain some further information about the regular representation γ. We first note that the character of γ is given by

$$\chi_\gamma(x) = he(x) = \begin{cases} h & x = \text{identity of } G, \\ 0 & \text{otherwise.} \end{cases}$$

Suppose now that $\alpha_1, \alpha_2, \ldots, \alpha_k$ are the pairwise inequivalent irreducible representations of G, where α_i is of degree m_i. Let v_i be the multiplicity of α_i in γ, so that

$$\gamma \sim v_1 \alpha_1 + v_2 \alpha_2 + \cdots + v_k \alpha_k,$$
$$\chi_\gamma = v_1 \chi_{\alpha_1} + v_2 \chi_{\alpha_2} + \cdots + v_k \chi_{\alpha_k}.$$

Then

$$m_1^2 + m_2^2 + \cdots + m_k^2 = h, \tag{26}$$
$$v_1 m_1 + v_2 m_2 + \cdots + v_k m_k = h, \tag{27}$$
$$v_1^2 + v_2^2 + \cdots + v_k^2 = h. \tag{28}$$

(26) is just theorem (15), (27) follows by comparing the degree of γ with the degrees of the α_i, and (28) is a consequence of (25). But now we find that

$$\sum_{i=1}^{k} (m_i - v_i)^2 = \sum_{i=1}^{k} (m_i^2 - 2m_i v_i + v_i^2) = h - 2h + h = 0,$$

so that

$$m_i = v_i, \quad 1 \le i \le k.$$

Hence we have proved

Theorem 18. *In the regular representation of a finite group* G, *each of the pairwise inequivalent irreducible representations of* G *occurs with multiplicity equal to its degree.*

We are going to determine the number of one-dimensional representations of G. For this purpose we introduce the important notion of *extending* a representation.

Let H be a normal subgroup of G, and

$$G = \sum x_i H$$

a coset decomposition of G modulo H. Let α be a representation of G/H. Then α can be extended to a representation of G as follows:

Let x be any element of G. Then $x \equiv x_i \bmod H$, for some i. Define

$$A(x) = A(x_i H), \quad x \in G.$$

It is a straightforward verification that this really defines a representation of G.

We now prove

Theorem 19. *The number of one-dimensional representations of the finite group* G *is just* $o(G/G') = (G : G')$, *where* G' *is the commutator subgroup of* G.

Proof. Let α be a one-dimensional representation of G/G'. Then α can be extended to a one-dimensional representation of G as above, and different representations give different extensions. Thus if N_1 denotes the number of one-dimensional representations of G, then $N_1 \geq (G : G')$, since the abelian group G/G' has exactly $o(G/G')$ one-dimensional representations, by theorem 8.

Now suppose that α is a one-dimensional representation of G. Then certainly

$$A(xy) = A(x)A(y) = A(y)A(x) = A(yx), \quad x, y \in G. \tag{29}$$

It follows easily from (29) that

$$A(x) = 1, \quad x \in G',$$

and thus that

$$A(x) = A(y), \quad x \equiv y \bmod G'.$$

Hence α certainly affords a one-dimensional representation of G/G', and so $N_1 \leq (G : G')$. Thus $N_1 = (G : G')$, and the proof of the theorem is concluded. □

The best way to think of G/G' is as G made abelian: that is, as G with the additional relation of commutativity added to it. This interpretation is particularly valuable when G is given in terms of generators and relations. For example, the dihedral group D_n is the group generated by x, y with defining relations $x^2 = (xy)^2 = y^n = 1$. Adding the relation of commutativity, we find that D_n/D_n' is the group generated by x, y with defining relations $x^2 = 1$, $y^{(n,2)} = 1$, $xy = yx$, and so is of order $2(n, 2)$.

In the examples that follow, we retain our previous notation and introduce the following one: Let G be a group of order h and with k conjugacy classes. Then the

number of pairwise inequivalent irreducible representations of degree d will be
denoted by N_d. By theorem 11, it suffices to restrict d to divisors of h. Then
theorems 15 and 16 become

$$h = \sum_{d|h} d^2 N_d, \tag{30}$$

$$k = \sum_{d|h} N_d; \tag{31}$$

and theorem 19 states that

$$N_1 = o(G/G'). \tag{32}$$

Of course not all d can occur. Formula (30) implies for example that d satisfies
the inequality

$$d \leq \sqrt{h}. \tag{33}$$

Examples

(a) Let G be the quaternion group of order 8. Then G may be defined as the
group generated by elements x, y with defining relations

$$x^4 = 1, \quad x^2 = y^2, \quad xy^{-1} = yx.$$

The elements of G may be exhibited as

$$x^p, x^p y, \quad p = 0, 1, 2, 3.$$

G has 5 conjugacy classes, and as representatives of the conjugacy classes we may
take

$$1, x, x^2, y, xy.$$

Since G/G' is the direct product of 2 cyclic groups of order 2, $N_1 = 4$. Furthermore
the equation

$$4 + 4N_2 = 8$$

implies that $N_2 = 1$.

The one-dimensional representations are given by

$$x \rightarrow 1, -1, \quad 1, -1$$
$$y \rightarrow 1, \quad 1, -1, -1$$

and the one of degree 2 may be taken as

$$x \to \begin{pmatrix} i & 0 \\ 0 & -i \end{pmatrix}, \quad y \to \begin{pmatrix} 0 & 1 \\ -1 & 0 \end{pmatrix}.$$

The character table is

	$[1]$	$[x]$	$[x^2]$	$[y]$	$[xy]$
	1	2	1	2	2
α_1	1	1	1	1	1
α_2	1	-1	1	1	-1
α_3	1	1	1	-1	-1
α_4	1	-1	1	-1	1
α_5	2	0	-2	0	0

The group G is better known as the group of quaternionic units $\pm 1, \pm i, \pm j, \pm k$, where multiplication satisfies

$$i^2 = j^2 = k^2 = ijk = -1.$$

(b) Let G be the tetrahedral group of order 12. Then G may be described as the group generated by elements x, y with defining relations

$$x^2 = y^3 = (xy)^3 = 1,$$

and has 4 conjugacy classes. Since G/G' is cyclic of order 3, $N_1 = 3$, and the 1-dimensional representations are

$$x \to 1, \ 1, \ 1$$
$$y \to 1, \ \rho, \ \rho^2,$$

where ρ is a primitive cube root of unity. Furthermore the equation

$$3 + 4N_2 + 9N_3 = 12$$

has the unique solution $N_2 = 0$, $N_3 = 1$. Hence, apart from the 1-dimensional representations, G has just one other irreducible representation, which is of degree 3. This is realizable as the group of spatial rotations which transform the regular tetrahedron into itself, and may be taken so that

$$x \to \begin{pmatrix} 1 & 0 & 0 \\ 0 & -1 & 0 \\ 0 & 0 & -1 \end{pmatrix}, \quad y \to \begin{pmatrix} 0 & 0 & 1 \\ 1 & 0 & 0 \\ 0 & 1 & 0 \end{pmatrix}.$$

(c) Let G be the octahedral group of order 24. Then G may be described as the group generated by elements x, y with defining relations

$$x^2 = y^3 = (xy)^4 = 1,$$

and has 5 conjugacy classes. Since G/G' is cyclic of order 2, $N_1 = 2$, and the 1-dimensional representations are

$$x \to 1, -1$$
$$y \to 1, \quad 1.$$

Furthermore the equation

$$2 + 4N_2 + 9N_3 + 16N_4 = 24$$

has the unique solution $N_2 = 1$, $N_3 = 2$, $N_4 = 0$. Hence there is one irreducible representation of degree 2 and two of degree 3, which may be taken as the orthogonal representations

$$x \to \begin{pmatrix} -1 & 0 \\ 0 & 1 \end{pmatrix}, \begin{pmatrix} -1 & 0 & 0 \\ 0 & 0 & 1 \\ 0 & 1 & 0 \end{pmatrix}, \begin{pmatrix} 1 & 0 & 0 \\ 0 & 0 & -1 \\ 0 & -1 & 0 \end{pmatrix},$$

$$y \to \begin{pmatrix} -1/2 & -1/2\sqrt{3} \\ 1/2\sqrt{3} & -1/2 \end{pmatrix} \begin{pmatrix} 0 & 0 & 1 \\ 1 & 0 & 0 \\ 0 & 1 & 0 \end{pmatrix}, \begin{pmatrix} 0 & 0 & 1 \\ 1 & 0 & 0 \\ 0 & 1 & 0 \end{pmatrix}.$$

G is realizable as the group of spatial rotations transforming the regular octahedron (or cube) into itself.

(d) Let G be the icosahedral group of order 60. Then G may be described as the group generated by elements x, y with defining relations

$$x^2 = y^3 = (xy)^5 = 1,$$

and has 5 conjugacy classes. Since G/G' is trivial, $N_1 = 1$, and

$$x \to 1$$
$$y \to 1$$

is the only 1-dimensional representation of G. Furthermore the equations

$$1 + 4N_2 + 9N_3 + 16N_4 + 25N_5 + 36N_6 = 60$$
$$1 + N_2 + N_3 + N_4 + N_5 + N_6 = 5$$

have the unique solution $N_2 = 0$, $N_3 = 2$, $N_4 = 1$, $N_5 = 1$, $N_6 = 0$. Thus there are two irreducible representations of degree 3, one of degree 4, and one of degree 5, which may be taken as the orthogonal representations

$$
x \rightarrow
\begin{bmatrix}
-1 & 0 & 0 \\
0 & -\dfrac{1}{4} & -\dfrac{1}{4}\sqrt{15} \\
0 & -\dfrac{1}{4}\sqrt{15} & \dfrac{1}{4}
\end{bmatrix}
\begin{bmatrix}
-1 & 0 & 0 \\
0 & -\dfrac{1}{4} & -\dfrac{1}{4}\sqrt{15} \\
0 & -\dfrac{1}{4}\sqrt{15} & \dfrac{1}{4}
\end{bmatrix},
$$

$$
y \rightarrow
\begin{bmatrix}
-\dfrac{1}{3} & -\dfrac{2}{3}\sqrt{2} & 0 \\
\dfrac{1}{3}\sqrt{2} & -\dfrac{1}{6} & \dfrac{1}{2}\sqrt{3} \\
-\dfrac{1}{3}\sqrt{6} & \dfrac{1}{6}\sqrt{3} & \dfrac{1}{2}
\end{bmatrix}
\begin{bmatrix}
-\dfrac{1}{3} & \dfrac{2}{3}\sqrt{2} & 0 \\
-\dfrac{1}{3}\sqrt{2} & -\dfrac{1}{6} & -\dfrac{1}{2}\sqrt{3} \\
-\dfrac{1}{3}\sqrt{6} & -\dfrac{1}{6}\sqrt{3} & \dfrac{1}{2}
\end{bmatrix},
$$

$$
x \rightarrow
\begin{bmatrix}
-\dfrac{1}{4} & \dfrac{1}{4}\sqrt{15} & 0 & 0 \\
\dfrac{1}{4}\sqrt{15} & \dfrac{1}{4} & 0 & 0 \\
0 & 0 & 1 & 0 \\
0 & 0 & 0 & -1
\end{bmatrix}
\begin{bmatrix}
1 & 0 & 0 & 0 & 0 \\
0 & -\dfrac{1}{2} & 0 & \dfrac{1}{2}\sqrt{3} & 0 \\
0 & 0 & \dfrac{1}{2} & 0 & -\dfrac{1}{2}\sqrt{3} \\
0 & \dfrac{1}{2}\sqrt{3} & 0 & \dfrac{1}{2} & 0 \\
0 & 0 & -\dfrac{1}{2}\sqrt{3} & 0 & -\dfrac{1}{2}
\end{bmatrix},
$$

$$
y \rightarrow
\begin{bmatrix}
1 & 0 & 0 & 0 \\
0 & -\dfrac{1}{3} & \dfrac{2}{3}\sqrt{2} & 0 \\
0 & -\dfrac{1}{3}\sqrt{2} & -\dfrac{1}{6} & -\dfrac{1}{2}\sqrt{3} \\
0 & -\dfrac{1}{3}\sqrt{6} & -\dfrac{1}{6}\sqrt{3} & \dfrac{1}{2}
\end{bmatrix}
\begin{bmatrix}
-\dfrac{1}{3} & \dfrac{2}{3}\sqrt{2} & 0 & 0 & 0 \\
-\dfrac{1}{3}\sqrt{2} & -\dfrac{1}{6} & -\dfrac{1}{2}\sqrt{3} & 0 & 0 \\
-\dfrac{1}{3}\sqrt{6} & -\dfrac{1}{6}\sqrt{3} & \dfrac{1}{2} & 0 & 0 \\
0 & 0 & 0 & -\dfrac{1}{2} & \dfrac{1}{2}\sqrt{3} \\
0 & 0 & 0 & -\dfrac{1}{2}\sqrt{3} & -\dfrac{1}{2}
\end{bmatrix}.
$$

G may be realized as the group of spatial rotations transforming the regular icosahedron (or dodecahedron) into itself.

(e) Let D_n denote the dihedral group of order $2n$. Then D_n is the group generated by elements x, y with defining relations

$$x^2 = (xy)^2 = y^n = 1.$$

Since $xy^r = y^{-r}x$, the elements of D_n may be exhibited as

$$y^p, xy^p, \quad p = 0, 1, \ldots, n-1.$$

The number of conjugacy classes of D_n is easily computed, and is given by

$$k = \begin{cases} \dfrac{n}{2} + \dfrac{3}{2} & n \text{ odd} \\[2mm] \dfrac{n}{2} + 3 & n \text{ even.} \end{cases}$$

As representatives of the conjugacy classes we may take

$$x, y^r, \qquad 0 \le r \le \frac{n-1}{2}, \quad (n \text{ odd}),$$

$$x, xy, y^r, \quad 0 \le r \le \frac{n}{2}, \qquad (n \text{ even}).$$

We first compute N_1. In any representation of D_n, we need only specify the homomorphic images of x and y. If $x \to \epsilon$, $y \to \zeta$ then

$$\epsilon^2 = (\epsilon\zeta)^2 = \zeta^n = 1.$$

This gives the following possibilities:

$$x \to 1, \quad -1 \quad (n \text{ odd}),$$
$$y \to 1, \quad 1$$
$$x \to 1, \quad -1, \quad 1, \quad -1 \quad (n \text{ even}).$$
$$y \to 1, \quad 1, \quad -1, \quad -1.$$

Thus

$$N_1 = \begin{cases} 2 & n \text{ odd} \\ 4 & n \text{ even.} \end{cases}$$

We now compute that

$$S = \sum_{\substack{d|2n \\ d>2}} (d^2 - 4)N_d = 2n - 4k + 3N_1,$$

and

$$2n - 4k + 3N_1 = \begin{cases} 2n - 2n - 6 + 6 = 0 & n \text{ odd} \\ 2n - 2n - 12 + 12 = 0 & n \text{ even.} \end{cases}$$

Thus $S = 0$. Since $d^2 - 4 > 0$ for $d > 2$, this implies that $N_d = 0$ for $d > 2$. Hence we have shown that the irreducible representations of D_n are all of degree 1 or 2.

We have $N_1 + 4N_2 = 2n$. Thus

$$N_2 = \frac{n}{2} - \frac{N_1}{4} = \begin{cases} \dfrac{n-1}{2} & n \text{ odd} \\ \dfrac{n}{2} - 1 & n \text{ even.} \end{cases}$$

The actual representations of degree 2 are easily computed. It turns out that they may be given by

$$x \to \begin{pmatrix} 1 & 0 \\ 0 & -1 \end{pmatrix}, \quad y \to \begin{bmatrix} \cos\dfrac{2\pi k}{n} & -\sin\dfrac{2\pi k}{n} \\ \sin\dfrac{2\pi k}{n} & \cos\dfrac{2\pi k}{n} \end{bmatrix},$$

where $1 \le k \le \left[\frac{n-1}{2}\right]$. The faithful ones are those for which $(k, n) = 1$. Alternatively, one can take

$$x \to \begin{pmatrix} 0 & 1 \\ 1 & 0 \end{pmatrix}, \quad y \to \begin{pmatrix} \zeta^k & 0 \\ 0 & \zeta^{-k} \end{pmatrix}, \quad \zeta = e^{\frac{2\pi i}{n}}, \quad 1 \le k \le \left[\frac{n-1}{2}\right].$$

(f) Let $\alpha = \{A(x) : x \in G\}$ be a representation of the group G, and let H be a subgroup of G. Then the *restriction* of α to H is the representation

$$\alpha^H = \{A(x) : x \in H\},$$

and is said to be *induced* by α. We denote the character of α^H by χ_α^H. We shall show that if G is finite, then every irreducible representation of H occurs as a component of some irreducible representation of G restricted to H.

Let $\alpha_1, \alpha_2, \ldots, \alpha_k$ be the pairwise inequivalent irreducible representations of G, and suppose that α_i is of degree m_i. Let $\beta_1, \beta_2, \ldots, \beta_l$ be the pairwise inequivalent irreducible representations of H, and suppose that β_j is of degree n_j. Since α_i^H is a representation of H, nonnegative integers v_{ij} exist such that

$$\alpha_i^H \sim v_{i1}\beta_1 + v_{i2}\beta_2 + \cdots + v_{il}\beta_l.$$

Thus

$$\chi_{\alpha_i}^H(x) = \sum_{j=1}^{l} v_{ij} \chi_{\beta_j}(x), \quad x \in H.$$

It follows from the orthogonality relationships (10) and (12) that

$$v_{ij} = \frac{1}{o(H)} \sum_{x \in H} \chi_{\alpha_i}^H(x) \chi_{\beta_j}(x^{-1}).$$

Multiply by m_i and sum over all i, using the relationship

$$\frac{1}{o(G)} \sum_{i=1}^{k} m_i \chi_{\alpha_i}(x) = e(x), \quad x \in G,$$

which is a consequence of the orthogonality relationship (24). Since $\chi_{\alpha_i}^{H}(x) = \chi_{\alpha_i}(x)$ for all $x \in H$, the result is that

$$\sum_{i=1}^{k} m_i v_{ij} = (G : H)n_j.$$

Hence v_{ij} cannot be zero for all i and a fixed j. This completes the proof.

(g) Let $\alpha = \{A(x) : x \in G\}$ be an irreducible representation of degree m of G. Let C be the center of G. Then if $z \in C$, $A(z) = \lambda_z I$, since $A(z)$ commutes with each matrix of α and α is irreducible.

Suppose now that G is finite, and set $(G : C) = p$. Let $G = \sum_{1 \leq i \leq p} x_i C$ be the coset decomposition of G modulo C, and take $x_1 = 1$. Then every element $x \in G$ has a unique expression as

$$x = x_i z, \quad 1 \leq i \leq p, \quad z \in C,$$

and

$$A(x) = A(x_i z) = A(x_i)A(z) = \lambda_z A(x_i).$$

Since $|\lambda_z| = 1$ and α is irreducible, it follows that

$$o(G) = \sum_{x \in G} |\chi_\alpha(x)|^2 = \sum_{\substack{1 \leq i \leq p \\ z \in C}} |\chi_\alpha(x_i)|^2 = o(C) \sum_{1 \leq i \leq p} |\chi_\alpha(x_i)|^2,$$

so that

$$\sum_{1 \leq i \leq p} |\chi_\alpha(x_i)|^2 = o(G)/o(C) = p.$$

Since $\chi_\alpha(x_1) = m$, this implies that

$$m^2 \leq (G : C).$$

As a matter of fact, $m | (G : G)$, and Ito has shown that m divides the index of every abelian normal subgroup of G.

17   The Symmetric Group

The group of permutations of n symbols is denoted by S_n, and is of order $n!$. We cannot discuss this group fully here, and must content ourselves with mentioning some important properties.

Two elements of S_n are in the same conjugacy class if and only if they have the same cycle structure. Thus a 1-1 correspondence exists between the classes and the ordered n-tuples

$$(a_1, a_2, \ldots, a_n), \quad a_i \geq 0, \quad \sum_{i=1}^{n} i a_i = n,$$

which is set up by saying that an element of S_n belongs to class $(a_1, a_2, \ldots, a_n)$ if and only if it has a_1 1-cycles, a_2 2-cycles, ..., a_n n-cycles. Thus the number of conjugacy classes is just the number of solutions in nonnegative integers a_i of the diophantine equation

$$\sum_{i=1}^{n} i a_i = n;$$

and this is $p(n)$, the number of partitions of n. The generating function of $p(n)$ is

$$\sum_{n=0}^{\infty} p(n) x^n = \prod_{n=1}^{\infty} (1 - x^n)^{-1},$$

where $p(0) = 1$.

The number of elements in the class $(a_1, a_2, \ldots, a_n)$ is

$$\frac{n!}{1^{a_1} a_1! 2^{a_2} a_2! \cdots n^{a_n} a_n!}.$$

The commutator subgroup S_n' of S_n is of index 2 for $n > 1$. Thus S_n has just 2 one-dimensional representations for $n > 1$: the principal representation, which assigns 1 to each element of S_n; and the alternating representation, which assigns 1 to elements of S_n belonging to S_n', and -1 to elements of S_n not belonging to S_n'. The group S_n' is just the alternating group A_n, and is usually characterized as the subgroup of S_n consisting of the even permutations; that is, those permutations which are products of an even number of transpositions. For $n > 4$ A_n is a simple group, and is the only proper normal subgroup of S_n. This has the consequence that for $n > 4$, every irreducible representation of S_n of degree > 1 is faithful. Furthermore every representation of S_n is equivalent to one by integral matrices.

If the partitions of n are exhibited as

$$n = b_1 + b_2 + \cdots + b_r,$$
$$b_1 \geq b_2 \geq \cdots \geq b_r,$$

then the irreducible representation associated with this partition has degree

$$d = \frac{n! \prod_{i<j} (c_i - c_j)}{c_1! c_2! \cdots c_r!},$$

where

$$c_i = b_i + r - i, \quad 1 \leq i \leq r,$$

and the product in the formula above is taken to be 1 when $r = 1$. If $n > 4$, then either $d = 1$ or $d \geq n - 1$, and the representations occur in associated pairs α, $\alpha\sigma$ (see sec. 19), where σ is the alternating representation. It is possible for α to be equivalent to $\alpha\sigma$.

Examples

(a) Choose $n = 5$. Then the partitions of n are

$$5, \ 4+1, \ 3+2, \ 3+1+1, \ 2-2+1, \ 2+1+1+1, \ 1+1+1+1+1.$$

For the degrees of the associated representations we find

partition	5	4+1	3+2	3+1+1	2+2+1	2+1+1+1	1+1+1+1+1
degree	1	4	5	6	5	4	1.

(b) Let $\pi = \{P\}$ be the totality of $n \times n$ permutation matrices. Then considered as a multiplicative group, π affords a representation of S_n of degree n.

Let $t(k)$ $(0 \leq k \leq n)$ denote the number of elements of π of trace k; or equivalently, the number of $n \times n$ permutation matrices having precisely k ones on the diagonal. Then $t(k) = \binom{n}{k} d(n-k)$, where $d(n-k)$ is the number of "derangements" of $n - k$ symbols; that is, the number of permutations of $n - k$ symbols leaving no symbol fixed. Counting elements of π by trace, we have

$$\sum_{k=0}^{n} \binom{n}{k} d(n-k) = n!,$$

so that

$$\sum_{k=0}^{n} \frac{1}{k!} \frac{d(n-k)}{(n-k)!} = 1.$$

Going over to the associated generating functions we get

$$e^x \sum_{n=0}^{\infty} \frac{d(n)}{n!} x^n = \frac{1}{1-x},$$

where we have defined $d(0)$ to be 1.

Now an easy calculation shows that

$$\sum_{n=0}^{\infty} \frac{n^2}{n!} x^n = (x^2 + x)e^x.$$

Hence

$$\sum_{n=0}^{\infty} \frac{n^2}{n!} x^n \cdot \sum_{n=0}^{\infty} \frac{d(n)}{n!} x^n = \frac{x^2 + x}{1-x} = x + \frac{2x^2}{1-x}.$$

This yields the identity

$$\sum_{k=0}^{n} \frac{k^2}{k!} \frac{d(n-k)}{(n-k)!} = 2, \quad n > 1,$$

so that

$$\frac{1}{n!} \sum_{k=0}^{n} \binom{n}{k} d(n-k)k^2 = 2, \quad n > 1.$$

This is equivalent to the relation

$$\frac{1}{n!} \sum_{P \in \pi} |\mathrm{tr}(P)|^2 = 2, \quad n > 1,$$

which implies that π is composed of just two irreducible components. Since one of them is one-dimensional (example c, p. 2), the other must be of degree $n - 1$. Hence we have proved the existence of an irreducible representation of S_n of degree $n - 1$ for every $n > 1$.

An interesting corollary is that the dimension of the vector space spanned by the $n \times n$ permutation matrices is just $1^2 + (n-1)^2 = n^2 - 2n + 2$.

Chapter V

Some Theorems of Burnside

18   Conjugacy Classes

We are going to prove some of Burnside's theorems on finite groups, for which we require some information about conjugacy classes. We retain our previous notation: The finite group G is of order h and has the k conjugacy classes $[y_i]$, $1 \leq i \leq k$. Furthermore let $h_i = h(y_i)$ be the number of elements in $[y_i]$. Then certainly

$$h = h_1 + h_2 + \cdots + h_k. \tag{34}$$

Let $[y_1]$ be the class consisting of the identity element of G alone. Then $h_1 = h(y_1) = 1$. (34) is known as the *class equation* and is the source of much of our information about finite groups.

We now consider subsets of G in which repetition is allowed. If $\mathscr{S}$ is such a set and x appears in $\mathscr{S}$ with multiplicity $v(x)$, we write

$$\mathscr{S} = \bigcup_{x \in G} v(x) \cdot x. \tag{35}$$

Then $\mathscr{S}$ is uniquely determined by the nonnegative integers $v(x)$.

A *normal set* $\mathscr{S}$ is one such that

$$x \mathscr{S} x^{-1} = \mathscr{S}, \quad x \in G.$$

It is readily verified that if $\mathscr{S}_1$, $\mathscr{S}_2$ are normal sets, then $\mathscr{S}_1 \mathscr{S}_2$ is a normal set and $\mathscr{S}_1 \mathscr{S}_2 = \mathscr{S}_2 \mathscr{S}_1$.

Let $\mathscr{S}$ be any subset of G, and suppose $\mathscr{S}$ given by (35). Let y be any element of G. Then

$$y\mathscr{S}y^{-1} = \bigcup_{x\in G} v(x)\cdot yxy^{-1} = \bigcup_{x\in G} v(y^{-1}xy)\cdot x.$$

Thus $\mathscr{S}$ is normal if and only if

$$v(x) = v(y^{-1}xy), \quad x, y \in G.$$

Hence if $\mathscr{S}$ is normal then elements in the same conjugacy class appear with equal multiplicities, and it follows that

$$\mathscr{S} = \bigcup_{i=1}^{k} v(y_i)[y_i].$$

Since the classes $[y_i]$ are themselves normal sets, we have the *class multiplication rule*

$$[y_i][y_j] = \bigcup_{p=1}^{k} v_{ijp}[y_p], \quad 1 \le i, j \le k, \tag{36}$$

where the v_{ijp} are nonnegative integers which satisfy $v_{ijp} = v_{jip}$. We shall use (36) shortly in connection with (16).

Now let $\mathscr{S}$ be an arbitrary subset of G. The *normalizer of* $\mathscr{S}$ is the subgroup of G consisting of all elements $x \in G$ such that $x\mathscr{S}x^{-1} = \mathscr{S}$, and will be denoted by $N(\mathscr{S})$.

We have the simple but important result:

Theorem 20. *The number τ of distinct sets*

$$x\mathscr{S}x^{-1}, \quad x \in G$$

is the index of $N(\mathscr{S})$.

Proof. Let

$$G = \sum_{i=1}^{t} x_i N(\mathscr{S})$$

be a coset decomposition for G modulo $N(\mathscr{S})$. Let x be any element of G. Then there is an i such that $1 \le i \le t$ and $x = x_i y$, where $y \in N(\mathscr{S})$. Thus

$$x\mathscr{S}x^{-1} = x_i y\mathscr{S}y^{-1}x_i^{-1} = x_i\mathscr{S}x_i^{-1},$$

so that $\tau \le t$. Furthermore these are distinct. For if

$$x_i\mathscr{S}x_i^{-1} = x_j\mathscr{S}x_j^{-1}, \quad 1 \le i, j \le t,$$

then $x_j^{-1}x_i \in N(\mathscr{S})$, which implies that $i = j$. Hence $\tau \ge t$, and so $\tau = t$. This completes the proof of the theorem. $\square$

It follows from this theorem that

$$h_i | h, \quad 1 \leq i \leq k, \tag{37}$$

since h_i is just the index of $N([y_i])$.

Suppose now that α is an irreducible representation of degree m of G. Then (16) states that

$$C_i = \sum_{x \in [y_i]} A(x) = \frac{h_i}{m} \chi_\alpha(y_i) \cdot I, \quad 1 \leq i \leq k. \tag{38}$$

Because of the class multiplication formula (36), we must have

$$C_i C_j = \sum_{p=1}^{k} v_{ijp} C_p, \quad 1 \leq i, j \leq k. \tag{39}$$

Then (38) and (39) imply that

$$\tau_i \tau_j = \sum_{p=1}^{k} v_{ijp} \tau_p, \quad 1 \leq i, j \leq k, \tag{40}$$

where

$$\tau_i = \frac{h_i}{m} \chi_\alpha(y_i), \quad 1 \leq i \leq k.$$

We may write (40) as

$$N_p \tau = \tau_p \tau, \quad 1 \leq p \leq k,$$

where N_p is the $k \times k$ matrix (v_{pij}) and $\tau = (\tau_1, \tau_2, \ldots, \tau_k)^T$. Since $\tau_1 = 1, \tau \neq 0$. It follows that τ_p is an eigenvalue of N_p with corresponding eigenvector τ. Since N_p is an integral matrix, theorem 12 of appendix A implies

Theorem 21. *The numbers*

$$\frac{h_i}{m} \chi_\alpha(y_i), \quad 1 \leq i \leq k$$

are algebraic integers.

At this point we require some additional information about algebraic integers. We prove

Lemma 6. *Suppose that the algebraic integer θ and all its conjugates are of modulus ≤ 1. Then θ is either 0 or a root of unity.*

Proof. Suppose that K, the algebraic number field to which θ belongs, is of degree n over the rationals. Assume that $\theta \neq 0$. Then if p is the degree of θ, $p \leq n$

(in fact p is a divisor of n, by theorem 7 of appendix A). Let $\theta_1 = \theta, \theta_2, \ldots, \theta_p$ be the conjugates of θ. Then if

$$f(x) = x^p - c_1 x^{p-1} + c_2 x^{p-2} - \cdots + (-1)^p c_p$$

is the monic irreducible polynomial with rational integral coefficients of which θ is a root, we have

$$|c_1| = |\theta_1 + \theta_2 + \cdots + \theta_p| \leq p,$$

$$|c_2| = |\theta_1 \theta_2 + \theta_1 \theta_3 + \cdots + \theta_{p-1} \theta_p| \leq \binom{p}{2},$$

$$\cdots$$

$$|c_p| = |\theta_1 \theta_2 \cdots \theta_p| \leq \binom{p}{p}.$$

It follows that the total number of polynomials $f(x)$ is bounded, and hence that K can contain only finitely many integers θ with the prescribed property.

Now consider the algebraic integer θ^r, where r is any positive rational integer. By theorem 7 of appendix A, the conjugates of θ are the different numbers among θ_i^r, $1 \leq i \leq p$. It follows that θ^r and all its conjugates are also of modulus ≤ 1. Hence the sequence

$$\theta, \theta^2, \ldots$$

can contain only finitely many different members. Thus rational integers r, s exist such that $0 < r < s$ and $\theta^r = \theta^s$, so that $\theta^{s-r} = 1$. This completes the proof of the lemma. $\qquad \square$

We use this lemma to prove

Lemma 7. *Suppose that* $(h_i, m) = 1$. *Then either* $\chi_\alpha(y_i) = 0$, *or* $A(y_i)$ *is scalar.*

Proof. Since $(h_i, m) = 1$, rational integers r, s exist such that $r h_i + s m = 1$. Hence $\chi_\alpha(y_i)/m = r h_i \chi_\alpha(y_i)/m + s \chi_\alpha(y_i)$, and since $h_i \chi_\alpha(y_i)/m$ and $\chi_\alpha(y_i)$ are algebraic integers, $\chi_\alpha(y_i)/m$ must also be an algebraic integer. But $\chi_\alpha(y_i)$ is the sum of m roots of unity, and so the same is true of all the conjugates of $\chi_\alpha(y_i)$. It follows that the algebraic integer $\chi_\alpha(y_i)/m$ and all its conjugates are of modulus ≤ 1. By lemma 6, this means that $\chi_\alpha(y_i)/m$ is either 0 or a root of unity. If $\chi_\alpha(y_i)/m$ is a root of unity, say ζ, then the eigenvalues of $A(y_i)$ must all be equal to ζ. Since $A(y_i)$ is diagonable, this implies that $A(y_i) = \zeta I$. This completes the proof of the lemma. $\qquad \square$

We can now prove:

Theorem 22 (Burnside). *Suppose that the number of elements in some conjugacy class of G is a prime power. Then G is not simple.*

Proof. Suppose that the pairwise inequivalent irreducible representations of G are $\alpha_1, \alpha_2, \ldots, \alpha_k$, where α_i is of degree m_i, and α_1 is the principal representation. Assume that h_l is a prime power, say p^t. Then $p|h$, and the relationship

$$h = 1 + m_2^2 + \cdots + m_k^2$$

implies that $(p, m_i) = 1$ for some i such that $2 \le i \le k$. From the formula for the character of the regular representation and theorem 16, we have (since y_l is not the identity)

$$1 + \sum_{i=2}^{k} m_i \chi_{\alpha_i}(y_l) = 0.$$

Thus it cannot happen that $\chi_{\alpha_i}(y_l) = 0$ whenever $(m_i, p) = 1$, $2 \le i \le k$. Thus there must be some i, $2 \le i \le k$, such that

$$(m_i, p) = 1, \quad \chi_{\alpha_i}(y_l) \ne 0.$$

If $m_i = 1$ then α_i is of degree 1 and different from the principal representation, which implies that G' is a proper normal subgroup of G. Suppose $m_i > 1$. Then lemma 7 implies that $A(y_l)$ is scalar. Hence the set of elements $y \in G$ such that $A(y)$ is scalar contains more than the identity alone, and cannot consist of all of G, since α_i is irreducible; and this set is certainly a normal subgroup of G. This completes the proof of the theorem.   □

Theorem 22 together with some standard results from group theory imply:

Theorem 23. *Every group G of order* $p^r q^s$, *where p, q are primes, is solvable.*

Proof. The class equation

$$p^r q^s = 1 + h_2 + \cdots + h_k$$

shows that pq cannot divide every h_i, $2 \le i \le k$. Thus there is an i, $2 \le i \le k$, such that $h_i = 1$ or h_i is a prime power. In the first case we conclude that the center of G contains more than the identity alone, and in the second case we conclude that G is not simple. In either case we have that G must contain a proper normal subgroup N. It is known that if N and G/N are solvable, then so is G. Using this fact the theorem readily follows by induction on the order of G.   □

Examples and Applications

The results obtained provide a good deal of information which can be used to advantage in the area of group theory. We give a number of examples.

(a) Let G have order p^2, p prime. Then the possibilities for m_i, $1 \leq i \leq k$, are just 1, p, or p^2. Since

$$\sum_{i=1}^{k} m_i^2 = p^2,$$

only 1 or p can occur. But every group has representations of degree 1, and so p cannot occur. Hence all m_i are 1, which implies that G is abelian.

(b) Let G have order p^n, p prime. Then

$$N_1 + p^2 N_p + P^4 N_{p^2} + \cdots = p^n,$$

so that $N_1 \equiv 0 \bmod p$. We have that $N_1 = (G : G')$, where G' is the commutator subgroup of G. Thus if $o(G') = p^{n_1}$, then $n_1 \leq n - 1$. We can repeat this argument with G' instead of G, and deduce that the sequence

$$G \supset G' \supset G'' \supset \cdots$$

must terminate. Hence we have proved that a group of prime power order is solvable, which can also be proved by theorem 22.

(c) Let G be nonabelian and of order p^3, p prime. Then $N_1 + p^2 N_p = p^3$, so that $N_1 \equiv 0 \bmod p^2$. Since N_1 divides p^3 and G is nonabelian, N_1 must be p^2. Thus $N_p = p - 1$, and the number of conjugacy classes k must be $k = N_1 + N_p = p^2 + p - 1$.

Suppose for example that G is the group generated by elements x, y with defining relations

$$x^{p^2} = y^p = 1, \quad y^{-1}xy = x^{p+1}.$$

Since G/G' is the direct product of two cyclic groups of order p, the p^2 one-dimensional representations of G are given by

$$x \to \xi, \quad y \to \eta$$

where ξ, η are any pth roots of unity.

We now consider the $p - 1$ irreducible representations of degree p. Let α be any such representation, and suppose that it is given by

$$x \to A, \quad y \to B,$$

where the $p \times p$ matrices A, B satisfy

$$A^{p^2} = B^p = I, \quad B^{-1}AB = A^{p+1}.$$

The relationship $y^{-1}xy = x^{p+1}$ implies that $y^{-1}x^p y = x^{p^2+p} = x^p$, so that x^p commutes with y. Since x^p also commutes with x, x^p is central. This implies that

A^p is scalar: $A^p = aI$, where $a^p = 1$ (since $A^{p^2} = I$). Now $a = 1$ implies that $A^p = I$, which implies that

$$AB = BA^{p+1} = BA,$$

so that A and B commute. Since α is irreducible and of degree $p > 1$, this is a contradiction. Hence $a \neq 1$, and the possible values for a are

$$a = \zeta^r, \quad 1 \leq r \leq p-1, \quad \zeta = \exp\left(\frac{2\pi i}{p}\right).$$

The eigenvalues of A may thus be exhibited as

$$\zeta^{r/p+t} (n_t \text{ times}), \quad 0 \leq t \leq p-1.$$

Then

$$n_0 + n_1 + \cdots + n_{p-1} = p,$$
$$n_0 + n_1\zeta + \cdots + n_{p-1}\zeta^{p-1} = 0$$

(since $B^{-1}AB = aA$ implies that A has zero trace). The latter equation and the fact that the cyclotomic polynomial $\frac{x^p-1}{x-1}$ is irreducible (see appendix B) imply that the n_i are all equal. The former equation then implies that they are all 1. Thus the eigenvalues of A are

$$\zeta^{r/p+t}, \quad 0 \leq t \leq p-1.$$

We can suppose (after a suitable conjugacy has been performed) that

$$A = \zeta^{r/p} \operatorname{diag}(1, \zeta, \ldots, \zeta^{p-1}).$$

Put $B = (b_{ij})$. Then the relationship $AB = aBA = \zeta^r BA$ implies that

$$(\zeta^{i-1} - \zeta^{r+j-1})b_{ij} = 0$$

Hence $b_{ij} = 0$ unless $i \equiv r + j \bmod p$. The matrix B therefore has just one nonzero entry in each row and column (and so is a generalized permutation matrix, or monomial matrix); and applying a suitable conjugacy by a diagonal matrix (which does not change A) we see that B may be taken as

$$B = \zeta^l \begin{pmatrix} 0 & \cdots & 0 & 1 \\ 1 & \cdots & 0 & 0 \\ & \cdots & & \\ 0 & \cdots & 1 & 0 \end{pmatrix}^r,$$

where $0 \leq l \leq p-1$.

Now $AB = \zeta^r BA$, so that for each integer s,

$$A^s BA^{-s} = \zeta^{rs} B.$$

Since $(r, p) = 1$, s may be determined so that $rs + l \equiv 0 \bmod p$. For this choice of s,

$$A^s BA^{-s} = \begin{pmatrix} 0 & \cdots & 0 & 1 \\ 1 & \cdots & 0 & 0 \\ & \cdots & & \\ 0 & \cdots & 1 & 0 \end{pmatrix}^r,$$

and of course

$$A^s AA^{-s} = A.$$

Hence we have shown that the $p - 1$ irreducible representations of degree p may be given by

$$x \to \zeta^{r/p} \operatorname{diag}(1, \zeta, \ldots, \zeta^{p-1}),$$

$$y \to \begin{pmatrix} 0 & \cdots & 0 & 1 \\ 1 & \cdots & 0 & 0 \\ & \cdots & & \\ 0 & \cdots & 1 & 0 \end{pmatrix}^r$$

where r may have any of the values $1, 2, \ldots, p - 1$.

When p is odd there is just one other nonabelian group of order p^3, namely the group G generated by three elements x, y, z with defining relations

$$x^p = y^p = z^p = 1, \quad xy = yxz, \quad xz = zx, \quad yz = zy.$$

Of course G may be generated by two elements alone, since z is expressible in terms of x and y; $z = x^{-1}y^{-1}xy$. Furthermore, the element z is central. It follows that in any irreducible representation of G, z must correspond to a scalar matrix, and x (say) may be taken diagonal.

An analysis similar to the preceding one shows that the p^2 one-dimensional representations are given by

$$x \to \xi, \quad y \to \eta, \quad z \to 1,$$

where ξ, η, are any pth roots of unity; and the $p-1$ irreducible representations of degree p may be taken as

$$x \to \mathrm{diag}(1, \zeta, \ldots, \zeta^{p-1}),$$

$$y \to \begin{pmatrix} 0 & \cdots & 0 & 1 \\ 1 & \cdots & 0 & 0 \\ & \cdots & & \\ 0 & \cdots & 1 & 0 \end{pmatrix}^r,$$

$$z \to \zeta^r I,$$

where r may have any of the values $1, 2, \ldots, p-1$.

When p is even there is also just one other nonabelian group of order $p^3 = 8$, namely the quaternion group. This has previously been discussed.

(d) Let G be a finite group of odd order $h = 2t+1$. We shall show that the only irreducible representation of G which is real is the principal representation, and that conjugate irreducible representations not equivalent to the principal representation are themselves inequivalent. (The principal representation is the representation of degree 1 such that every group element is assigned the value 1.)

Let α be any irreducible representation of G not equivalent to the principal representation α_1. Let m be the degree of α. Then m is odd, since $m \mid h$. Since h is odd, the group elements may be written as

$$x_0 = 1, x_1, x_1^{-1}, x_2, x_2^{-1}, \ldots, x_t, x_t^{-1}.$$

Then the orthogonality relationship

$$\sum_{x \in G} \chi_\alpha(x) \chi_{\alpha_1}(x^{-1}) = 0$$

becomes

$$m + \sum_{p=1}^{t} \{\chi_\alpha(x_p) + \chi_\alpha(x_p^{-1})\} = 0,$$

or

$$m + \sum_{p=1}^{t} \{\chi_\alpha(x_p) + \overline{\chi_\alpha(x_p)}\} = 0.$$

If χ_α is real, this implies that

$$\sum_{p=1}^{t} \chi_\alpha(x_p) = -\frac{m}{2}.$$

But the left side is an algebraic integer. This implies that m is even, a contradiction.

We now show that $\alpha \not\sim \bar{\alpha}$. For $\alpha \sim \bar{\alpha}$ implies that $\chi_\alpha = \chi_{\bar{\alpha}} = \overline{\chi_\alpha}$. Hence χ_α is real, and the previous argument implies that $\alpha \sim \alpha_1$, a contradiction.

These results tell us that among the pairwise inequivalent irreducible representations of G, the only real representation occurring is the principal one and the others occur in conjugate complex pairs. Thus the relationship

$$\sum_{i=1}^{k} m_i^2 = h$$

takes the form

$$1 + 2 \sum_{i=1}^{(k-1)/2} n_i^2 = h.$$

Since $n_i^2 \equiv 1 \bmod 8$, we obtain the curious result that $h \equiv k \bmod 16$, when h is odd.

(e) Let G have order pq, $p < q$, p, q primes. Then the only possibilities for the m_i are 1 or p. As before, let N_d denote the number of inequivalent irreducible representations of G of degree d. Then $N_1 + p^2 N_p = pq$. Thus $N_1 \equiv 0 \bmod p$. Since $N_1 = o(G)/o(G') = pq/o(G')$, we must have $o(G') = 1$ or q. We have shown therefore that G' must be abelian, so that $G'' = \{1\}$. Furthermore, the only possibilities for N_1 are $N_1 = p$ or pq. If $N_1 = pq$ then $N_p = 0$ and G is abelian, and therefore cyclic (since it contains an element of order p, an element of order q, and thus an element of order pq). If $N_1 = p$, then $N_p = \frac{q-1}{p}$, and thus $q \equiv 1 \bmod p$. We can conclude that there is just one group of order pq, namely the cyclic group of that order, if p, q are primes such that $p < q, q \not\equiv 1 \bmod p$.

There is only one non-abelian group G of order pq, which may be taken as the group generated by two elements x, y with defining relations

$$x^p = y^q = 1, \quad x^{-1}yx = y^r,$$

where $q = 1 + pt$, $r = g^t$, and g is a primitive root modulo q. The p one-dimensional representations are given by

$$x \to \xi, \quad y \to 1,$$

where ξ is any pth root of unity.

Let $x \to A, y \to B$, be an irreducible representation of degree p. Then

$$A^p = B^q = I, \quad A^{-1}BA = B^r.$$

As in example (c), we may choose B diagonal. We notice that the eigenvalues of B are all qth roots of unity, and that if θ is an eigenvalue of B, then so is θ^r. It follows that B may be chosen as

$$B = \operatorname{diag}(\theta, \theta^r, \ldots, \theta^{r^{p-1}}),$$

where θ is some primitive qth root of unity. But now the relationship $BA = AB^r$ implies that A is monomial, having nonzero entries only in positions $(1, p)$,

(2, 1), ..., (p, p − 1); and transforming by a diagonal matrix (which does not change B) we may in fact assume that

$$
A = \begin{pmatrix} 0 & \cdots & 0 & \lambda \\ 1 & \cdots & 0 & 0 \\ & \cdots & & \\ 0 & \cdots & 1 & 0 \end{pmatrix}.
$$

But $A^p = \lambda I$, and also $A^p = I$. Thus $\lambda = 1$, and it follows that the t irreducible representations of degree p may be taken as

$$
x \to \begin{pmatrix} 0 & \cdots & 0 & 1 \\ 1 & \cdots & 0 & 0 \\ & \cdots & & \\ 0 & \cdots & 1 & 0 \end{pmatrix},
$$

$$
y \to \operatorname{diag}(\zeta^s, \zeta^{rs}, \ldots, \zeta^{r^{p-1}s}),
$$

where $\zeta = e^{\frac{2\pi i}{q}}$ and s may have any of the t values $1, g, \ldots, g^{t-1}$.

(f) A group G whose Sylow subgroups are cyclic is said to be *S-metacyclic*. Thus a group of square-free order is certainly *S*-metacyclic. It is known that such a group G can always be generated by two elements x, y which satisfy the relations

$$
x^m = y^n = 1, \quad y^{-1}xy = x^r,
$$

where

$$
(m, n) = 1, \quad mn = h = \mathrm{o}(G), \quad r^n \equiv 1 \bmod m, \quad \text{and} \quad (r - 1, h) = 1.
$$

We will show that any irreducible representation of G is equivalent to a monomial representation (one in which the representing matrices have precisely one nonzero entry in each row and column) and that the degree of such a representation must be a divisor of n.

The one-dimensional representations are easily obtained, since G/G' is cyclic of order n, and are given by

$$
x \to 1, \quad y \to \eta,
$$

where η is any nth root of unity.

Suppose now that α is an irreducible representation of degree $t > 1$, and is given by

$$
x \to A, \quad y \to B,
$$

where the $t \times t$ matrices A, B satisfy

$$A^m = B^n = I, \quad B^{-1}AB = A^r.$$

Since A is of finite period, we may assume A diagonal. Furthermore, since A and A^r have the same eigenvalues, we may assume that

$$A = \sum_{s=1}^{l} \cdot (\zeta_s I_{p_s} \dotplus \zeta_s^r I_{p_s} \dotplus \cdots \dotplus \zeta_s^{r^{e_s-1}} I_{p_s}),$$

where ζ_s is an mth root of unity, e_s is the least positive integer such that $\zeta_s^{r^{e_s}} = \zeta_s$, and all the numbers

$$\zeta_s^{r^i}, \quad 0 \le i \le e_s - 1, \ 1 \le s \le l,$$

are distinct. Also, $\sum_{s=1}^{l} e_s p_s = t$.
    Set

$$Q_s = \begin{pmatrix} 0 & I_{p_s} & 0 & \cdots & 0 \\ 0 & 0 & I_{p_s} & \cdots & 0 \\ & & \cdots & & \\ I_{p_s} & 0 & 0 & \cdots & 0 \end{pmatrix}, \quad 1 \le s \le l,$$

$$Q = \sum_{s=1}^{l} \cdot Q_s.$$

Then $A^r = QAQ^T$, which implies that BQ commutes with A. This in turn implies that

$$BQ = \sum_{s=1}^{l} \cdot B_s,$$

$$B = \sum_{s=1}^{l} \cdot B_s Q_s^T.$$

It is clear that l must be 1 in order for α to be irreducible. Thus we have

$$A = \zeta I_p \dotplus \zeta^r I_p \dotplus \cdots \dotplus \zeta^{r^{e-1}} I_p,$$

where e is the least positive integer such that $\zeta^{r^e} = \zeta$, ζ is an mth root of unity, and $ep = t$.

Since BQ commutes with A, we must have

$$BQ = C_0 \dotplus C_1 \dotplus \cdots \dotplus C_{e-1},$$

$$B = \begin{pmatrix} 0 & \cdots & 0 & C_0 \\ C_1 & \cdots & 0 & 0 \\ 0 & \cdots & C_{e-1} & 0 \end{pmatrix}.$$

We can now determine a matrix

$$S = S_0 \dotplus S_1 \dotplus \cdots \dotplus S_{e-1}$$

such that

$$S^{-1}BS = \begin{pmatrix} 0 & \cdots & 0 & T \\ I_p & \cdots & 0 & 0 \\ & & \cdots & \\ 0 & \cdots & I_p & 0 \end{pmatrix},$$

and of course $S^{-1}AS = A$. Thus we may take

$$A = \zeta I_p \dotplus \zeta^r I_p \dotplus \cdots \dotplus \zeta^{r^{e-1}} I_p,$$

$$B = \begin{pmatrix} 0 & \cdots & 0 & T \\ I_p & \cdots & 0 & 0 \\ & & \cdots & \\ 0 & \cdots & I_p & 0 \end{pmatrix}.$$

Now

$$B^e = T \dotplus T \dotplus \cdots \dotplus T,$$

so that B^e commutes with A. Since B^e also commutes with B, B^e is central. This implies that $T = \lambda I_p$, where λ is some nth root of unity. A simple calculation with traces and an application of theorem 17 now show that p must be 1 in order for α to be irreducible. Thus

$$A = \operatorname{diag}(\zeta, \zeta^r, \ldots, \zeta^{r^{t-1}}),$$

$$B = \begin{pmatrix} 0 & \cdots & 0 & \lambda \\ 1 & \cdots & 0 & 0 \\ & & \cdots & \\ 0 & \cdots & 1 & 0 \end{pmatrix},$$

where ζ is an mth root of unity, t is the least positive integer such that $\zeta^{r^t} = \zeta$, and λ is an nth root of unity. Finally the facts that $B^n = I$, and B^s has elements on the diagonal if and only if s is divisible by t, imply that t divides n.

Chapter VI

Multiplication of Representations

19 Kronecker Products

The next topic that we treat is the important one of the multiplication of representations. If $A = (a_{ij})$, $B = (b_{ij})$ are arbitrary matrices, then the Kronecker product of A and B, written $A \otimes B$, is the matrix

$$A \otimes B = (a_{ij} B).$$

It is readily verified that

$$(A_1 \otimes B_1)(A_2 \otimes B_2) = (A_1 A_2) \otimes (B_1 B_2), \qquad (41)$$

provided only that the pairs A_1, A_2 and B_1, B_2 are conformal; i.e., may be multiplied in the order shown. Also if A and B are nonsingular then so is $A \otimes B$.

Suppose now that $\alpha = \{A(x) : x \in G\}$, $\beta = \{B(x) : x \in G\}$ are representations of the group G. We define the product of α and β by

$$\alpha\beta = \{A(x) \otimes B(x) : x \in G\}.$$

Then (41) implies that this is indeed a representation of G. Also if α is of degree m and β of degree n, then $\alpha\beta$ is of degree mn. Furthermore,

$$\chi_{\alpha\beta} = \chi_\alpha \chi_\beta. \qquad (42)$$

Formula (42) follows directly from the definition of the Kronecker product by computing traces, and allows us to establish multiplication rules for characters, as follows:

Let $\alpha_1, \alpha_2, \ldots, \alpha_k$ by the pairwise inequivalent irreducible representations of G. Then $\alpha_i \alpha_j$ is also a representation of G, and so nonnegative integers v_{ijp} exist such that for $1 \le i, j \le k$,

$$\alpha_i \alpha_j \sim v_{ij1}\alpha_1 + v_{ij2}\alpha_2 + \cdots + v_{ijk}\alpha_k.$$

It follows that

$$\chi_{\alpha_i}\chi_{\alpha_j} = \sum_{p=1}^{k} v_{ijp}\chi_{\alpha_p}, \quad 1 \le 1, j \le k. \tag{43}$$

We can in fact use (43) to define an algebra over the characters of dimension k for which the χ_{α_i}, $1 \le i \le k$, form a basis with structure constants

$$v_{ijp}, \quad 1 \le i, j, p \le k.$$

We cannot pursue this point further here.

We require the following lemma:

Lemma 8. *Let* $\alpha = \{A(x) : x \in G\}$ *be an irreducible representation of the finite group G. Then*

$$\sum_{x \in G} \chi_\alpha(x) = \begin{cases} 0 & \alpha \nsim \alpha_1 \\ h & \alpha \sim \alpha_1 \end{cases}. \tag{44}$$

Here α_1 is the principal representation; that is, the representation of degree 1 which assigns the value 1 to each element of the group.

Proof. Set

$$V = \sum_{x \in G} A(x).$$

Then V satisfies

$$A(y)V = VA(y) = V, \quad y \in G. \tag{45}$$

Since V commutes with each matrix of α, corollary 1 implies that V is scalar. Set $V = \lambda I$. Then (45) implies that

$$\lambda(A(y) - I) = 0, \quad y \in G,$$

from which we easily obtain

$$V = \begin{cases} 0 & \alpha \nsim \alpha_1 \\ h(1) & \alpha \sim \alpha_1. \end{cases} \qquad \square$$

The lemma follows upon taking traces.

Formula (44) can also be derived from (10) by choosing β as the principal representation.

Lemma 8 and the multiplication law (43) now imply:

Lemma 9. *Let α be a representation of the finite group* G, *and let* p(x) *be a polynomial with integral coefficients. Then*

$$\sum_{x \in G} p(\chi_\alpha(x)) \equiv 0 \ mod \ \text{h}.$$

Proof. Suppose that the pairwise inequivalent irreducible representations of G are $\alpha_1 \alpha_2, \ldots, \alpha_k$ and let α_i enter into α with multiplicity v_i. Then

$$\chi_\alpha = \sum_{i=1}^{k} v_i \chi_{\alpha_i},$$

and formula (43) implies that nonnegative integers μ_{ip} exist such that for each nonnegative integer p,

$$\chi_\alpha^p = \sum_{i=1}^{k} \mu_{ip} \chi_{\alpha_i}.$$

Hence

$$\sum_{x \in G} \chi_\alpha^p(x) = \sum_{i=1}^{k} \mu_{ip} \sum_{x \in G} \chi_{\alpha_i}(x), \tag{46}$$

and thus is divisible by h by lemma 8. The proof of the lemma now follows easily from (46). $\square$

We shall use this lemma shortly to derive some interesting facts about finite groups of integral matrices.

Let α be any representation of G of degree n, and let

$$t_1 = n, t_2, t_2, \ldots, t_r$$

be the distinct values assumed by $\chi_\alpha(x); x \in G$. Let ρ be the order of the kernel of α. Then

$$\sum_{x \in G} \{\chi_\alpha(x) - t_2\}\{\chi_\alpha(x) - t_3\} \cdots \{\chi_\alpha(x) - t_r\} = \rho(n - t_2)(n - t_3) \cdots (n - t_r),$$

since the sum is different from 0 if and only if $\chi_\alpha(x) = n$; i.e., if and only if $A(x) = I$. If we choose G to be any finite group of integral matrices and α the representation which assigns an element of G to itself, then the kernel of α is $\{I\}$ and lemma 8 implies

Theorem 24. *Let* G *be a finite group of integral* n × n *matrices. Let* $t_1 = n$, t_2, $t_3, \ldots, t_r$ *be the distinct values assumed by the traces of the elements of* G. *Then the order of* G *divides* $(n - t_2)(n - t_3) \cdots (n - t_r)$.

It follows for example that the order of G must be a devisor of $(2n)!$, since each t_i is the sum of n roots of unity, and consequently satisfies $|t_i| \leq n$.

As an application of this theorem, let us compute the finite subgroups of $SL(2, Z)$ (the multiplicative group of 2×2 rational integral matrices of determinant 1). Let $A \in SL(2, Z)$. Then the eigenvalues of A are reciprocals of one another (since det $A = 1$) and are algebraic integers of degree ≤ 2. If we suppose that A is of finite order, then we know by corollary 8 that a nonsingular matrix S exists such that

$$S^{-1}AS = \begin{pmatrix} \lambda & 0 \\ 0 & 1/\lambda \end{pmatrix}.$$

It follows that r is the order of A if and only if it is the order of λ. Thus λ must be a primitive rth root of unity, and so is of degree $\varphi(r)$ (see appendix B). Hence $\varphi(r) \leq 2$, which implies that $r = 1, 2, 3, 4, 6$. This limits λ to the values ± 1, $\pm i$, $\pm \rho$, $\pm \rho^2$, where ρ is a primitive cube root of unity, and so the trace of A can only be $0, \pm 1, \pm 2$. It follows that the order of any finite subgroup of $SL(2, Z)$ must divide

$$(2 - 0)(2 - 1)(2 + 1)(2 + 2) = 24.$$

In fact the only possible orders are 1, 2, 3, 4, 6, and any such subgroup must be conjugate over $GL(2, Z)$ (the multiplicative group of 2×2 rational integral matrices of determinant ± 1) to one of the cyclic groups

$$\{I\}, \{-I\}, \{U\}, \{T\}, \{-U\},$$

where $T = \begin{pmatrix} 0 & 1 \\ -1 & 0 \end{pmatrix}$, $U = \begin{pmatrix} 0 & 1 \\ -1 & -1 \end{pmatrix}$, and $T^2 = -I$, $U^3 = I$.

The situation is only slightly more complicated for $GL(2, Z)$. For if G is any integral matrix group, and G_1 the subgroup of G consisting of all elements of G of determinant 1, then G_1 is a normal subgroup of G of index 1 or 2. Since all the finite subgroups of $SL(2, Z)$ are C_1, C_2, C_3, C_4, C_6, and since G contains C_n as a normal subgroup of index 2 if and only if it is C_{2n}, $C_2 \times C_n$, or D_{2n}, a brief analysis shows that it is only necessary to add the groups D_1, D_2, D_3, D_4, D_6 to the above list to obtain all finite subgroups of $GL(2, Z)$. The number of new groups so obtained which are not conjugate over $GL(2, Z)$ is 8, and in terms of the canonical generators x, y of D_n (which satisfy $x^2 = (xy)^2 = y^n = 1$), a complete list is the following:

$$D_1: x \to \begin{pmatrix} 1 & 0 \\ 0 & -1 \end{pmatrix},$$

$$D_1': x \to \begin{pmatrix} 0 & 1 \\ 1 & 0 \end{pmatrix},$$

$$D_2 \colon x \to \begin{pmatrix} 1 & 0 \\ 0 & -1 \end{pmatrix}, \quad y \to -I,$$

$$D_2' \colon x \to \begin{pmatrix} 0 & 1 \\ 1 & 0 \end{pmatrix}, \quad y \to -I,$$

$$D_3 \colon x \to \begin{pmatrix} 0 & 1 \\ 1 & 0 \end{pmatrix}, \quad y \to \begin{pmatrix} 0 & 1 \\ -1 & -1 \end{pmatrix},$$

$$D_3' \colon x \to \begin{pmatrix} 0 & -1 \\ -1 & 0 \end{pmatrix}, \quad y \to \begin{pmatrix} 0 & 1 \\ -1 & -1 \end{pmatrix},$$

$$D_4 \colon x \to \begin{pmatrix} 0 & 1 \\ 1 & 0 \end{pmatrix}, \quad y \to \begin{pmatrix} 0 & 1 \\ -1 & 0 \end{pmatrix},$$

$$D_6 \colon x \to \begin{pmatrix} 0 & 1 \\ 1 & 0 \end{pmatrix}, \quad y \to \begin{pmatrix} 0 & 1 \\ -1 & 1 \end{pmatrix}.$$

Thus there are in all 13 unimodularly inequivalent 2×2 integral matrix groups. These lead in a natural way to the 17 planar crystallographic groups, which are concerned with the rigid motions in the plane (rotations, translations, reflections) leaving a given planar configuration invariant.

A similar discussion shows that the order of a finite subgroup of $SL(3, Z)$ must also be a divisor of 24, and that an individual element of such a group must have order 1, 2, 3, 4, 6. A careful enumeration shows that the possible groups are

(a) cyclic groups of orders 1, 2, 3, 4, 6,

(b) dihedral groups of orders 4, 6, 8, 12,

(c) the tetrahedral group of order 12,

(d) the octahedral group of order 24.

It can be shown that there are in all 70 unimodularly inequivalent 3×3 integral matrix groups, leading to 230 spatial crystallographic groups.

The maximal finite groups of 4×4 integral matrices have been determined by Dade, up to conjugacy. There are 9 of them and they can be described in terms of positive definite quadratic forms of which they are the automorph groups. See Dade's paper [10] for a complete discussion.

20 Powers of a Representation

Let α^p denote the pth Kronecker power of the representation α. In our discussion of the regular representation, we encountered a method for determining all irreducible representations of a finite group. The pth Kronecker power provides another such method. Specifically, we have:

Theorem 25. *Let α be a faithful representation of the finite group G. Then every irreducible representation of G occurs as a component of some Kronecker power of α.*

Proof. Let $\alpha_1, \alpha_2, \ldots, \alpha_k$ be the pairwise inequivalent irreducible representations of G, and suppose that α_i enters into α with multiplicity ν_i. Let α_i be of degree m_i, α of degree m. Then

$$\chi_\alpha = \sum_{i=1}^k \nu_i \chi_{\alpha_i},$$

$$\chi_{\alpha^p} = \chi_\alpha^p = \sum_{i=1}^k \nu_{ip} \chi_{\alpha_i},$$

where the ν_{ip} are nonnegative integers. Thus for each $x \in G$ we have

$$\chi_\alpha^p(x) = \sum_{i=1}^k \nu_{ip} \chi_{\alpha_i}(x).$$

If we multiply by $\chi_{\alpha_j}(x^{-1})$, sum over all $x \in G$, and make use of the orthogonality relationships (10) and (12), we obtain

$$h\nu_{jp} = \sum_{x \in G} \chi_\alpha^p(x) \chi_{\alpha_j}(x^{-1}).$$

Now consider the function

$$f_j(Z) = \sum_{p=1}^\infty \nu_{jp} Z^p.$$

Then $f_j(Z)$ is a rational function, and

$$hf_j(Z) = \sum_{x \in G} \chi_{\alpha_j}(x^{-1}) \frac{\chi_\alpha(x)Z}{1 - \chi_\alpha(x)Z}$$

$$= \frac{m_j m Z}{1 - m Z} + \cdots.$$

Since α is faithful, the denominator $1 - mZ$ cannot occur in any of the remaining terms. It follows that $f_j(Z)$ is not identically zero, and hence that ν_{jp} must be different from zero for at least one value of p. This completes the proof. □

Examples

(a) Let α, β, γ be representations of G. Then

$$(\alpha + \beta)\gamma = \alpha\gamma + \beta\gamma,$$
$$\gamma(\alpha + \beta) = \gamma\alpha + \gamma\beta.$$

(b) Let α, β be representations of the finite group G. Then if α contains p irreducible components and β q irreducible components, $\alpha\beta$ contains at least pq irreducible components. It follows that if $\alpha\beta$ is irreducible, then both α and β are irreducible.

(c) Let α, β be irreducible representations of the finite group G, and suppose that β is of degree one. Then $\alpha\beta$ is irreducible. For

$$\frac{1}{h}\sum_{x \in G}|\chi_{\alpha\beta}(x)|^2 = \frac{1}{h}\sum_{x \in G}|\chi_\alpha(x)\chi_\beta(x)|^2 = \frac{1}{h}\sum_{x \in G}|\chi_\alpha(x)|^2 = 1,$$

since $|\chi_\beta(x)| = 1$ for all $x \in G$.

21 Direct Products

The construction of section 19 can be modified slightly to give information on representations of direct products of groups. Let G, H be groups,

$$\alpha = \{A(x) : x \in G\}, \quad \beta = \{B(y) : y \in H\}$$

representations of G, H of respective degrees m, n. Then

$$\alpha \times \beta = \{A(x) \otimes B(y) : x \in G, y \in H\} \qquad (47)$$

is a representation of $G \times H$ of degree mn.

Conversely, it is not difficult to show that any representation of $G \times H$ is equivalent to the sum of representations of type (47). However, we can show much more. We denote the number of conjugacy classes of any group G by $k(G)$. A first result is the following:

Lemma 10. *Let* G, H *be finite groups. Then*

$$k(G \times H) = k(G)k(H).$$

Proof. Let

$$G = \sum_{i=1}^{r} C_i \, (r = k(G)), \quad H = \sum_{j=1}^{s} D_j \, (s = k(H)),$$

be the respective decompositions of G and H into their conjugacy classes. Then

$$G \times H = \sum_{\substack{i \leq i \leq r \\ 1 \leq j \leq s}} C_i \times D_j$$

is the decomposition of $G \times H$ into its conjugacy classes. For if C_i consists of all elements of G conjugate over G to x_i, and D_j consists of all elements of H conjugate over H to y_j, then $C_i \times D_j$ consists of all elements of $G \times H$ conjugate over $G \times H$ to $x_i \times y_j$. Furthermore the sets $C_i \times D_j$ are disjoint, no element of one set is conjugate to an element of another, and every element of $G \times H$ occurs in one of the sets. This completes the proof. $\qquad\qquad\square$

Lemma 11. *Let* G, H *be finite groups,* α *a representation of* G, β *a representation of* H. *Then* $\alpha \times \beta$ *is irreducible if and only if* α *and* β *are irreducible.*

Proof. Note that if $z = x \times y \in G \times H$, then $\chi_{\alpha \times \beta}(z) = \chi_\alpha(x)\chi_\beta(y)$; and that $o(G \times H) = o(G)o(H)$. These easily imply that

$$\frac{1}{o(G \times H)} \sum_{z \in G \times H} |\chi_{\alpha \times \beta}(z)|^2 = \left\{ \frac{1}{o(G)} \sum_{x \in G} |\chi_\alpha(x)|^2 \right\} \left\{ \frac{1}{o(H)} \sum_{y \in H} |\chi_\beta(y)|^2 \right\}.$$

$$\square$$

The lemma follows from this identity by applying formula (25), since the product of two positive integers is 1 if an only if each of them is 1.

We now prove:

Lemma 12. *Let* G, H *be finite groups. Then every irreducible representation* γ *of* G $\times$ H *is of the form* $\alpha \times \beta$, *where* α *is an irreducible representation of* G *and* β *an irreducible representation of* H.

Proof. Suppose that

$$\gamma = \{C(x \times y) : x \in G, y \in H\}.$$

Put

$$C(x \times 1) = A(x), \quad C(1 \times y) = B(y).$$

Then $\alpha = \{A(x) : x \in G\}$ is a representation of G, $\beta = \{B(y) : y \in H\}$ is a representation of H, and the matrices of the representations satisfy

$$A(x)B(y) = B(y)A(x), \quad x \in G, \ y \in H. \tag{48}$$

We can assume that α is reduced, so that

$$\alpha = v_1\alpha_1 + v_2\alpha_2 + \cdots + v_l\alpha_l,$$

where the representations $\alpha_1, \alpha_2, \ldots, \alpha_l$ are pairwise inequivalent and irreducible. Then (48) easily implies that $l = 1$, since γ is irreducible. Thus $\alpha = \nu_1\alpha_1 = \alpha_1 + \alpha_1 + \cdots + \alpha_1$. Writing

$$A(x) = A_1(x) \dot{+} A_1(x) \dot{+} \cdots \dot{+} A_1(x), \quad x \in G,$$
$$B(y) = (B_{ij}(y)), \quad y \in H,$$

(48) implies that

$$A_1(x)B_{ij}(y) = B_{ij}(y)A_1(x), \quad x \in G, \ y \in H.$$

Hence $B_{ij}(y)$ is scalar, since α_1 is irreducible:

$$B_{ij}(y) = \lambda_{ij}(y) \cdot I.$$

Put

$$(\lambda_{ij}(y)) = B_1(y), \quad y \in H.$$

Then the matrices $B_1(y)$ also form a representation of H, and $B(y) = B_1(y) \otimes I$. Since $A(x) = I \otimes A_1(x)$, we get

$$C(x \times y) = A(x)B(y) = (I \otimes A_1(x))(B_1(y) \otimes I)$$
$$= B_1(y) \otimes A_1(x), \quad x \in G, \ y \in H.$$

Thus $\gamma = \beta \times \alpha$, where $\beta = \{B_1(y) : y \in H\}, \alpha = \{A_1(x) : x \in G\}$. The conclusion now follows, since $\beta \times \alpha \sim \alpha \times \beta$.

Putting these together, we arrive at the following result:

Theorem 26. *Let* G, H *be finite groups. Suppose that* $\alpha_1, \alpha_2, \ldots, \alpha_r$ *are the pairwise inequivalent irreducible representations of* G, *and* $\beta_1, \beta_2, \ldots, \beta_s$ *the pairwise inequivalent irreducible representations of* H. *Then*

$$\alpha_i \times \beta_j, \quad 1 \le i \le r, \ 1 \le j \le s,$$

are the pairwise inequivalent irreducible representations of G $\times$ H.

Examples

(a) There are three non-abelian groups of order 12: the tetrahedral group, D_6, and $D_3 \times C_2$. The first two have already been discussed. We know that D_3 has 3 classes and irreducible representations of degrees 1, 1, 2; and C_2 has 2 classes and irreducible representations of degrees 1, 1. Hence $D_3 \times C_2$ has 6 classes and irreducible representations of degrees 1, 1, 1, 1, 2, 2. We can present D_3 as $\{x, y\}$, where $x^2 = (xy)^2 = y^3 = 1$, and C_2 as $\{z\}$, where $z^2 = 1$. The irreducible representations of D_3 are given by

$$x \to 1, \quad -1, \quad \begin{pmatrix} 0 & 1 \\ 1 & 0 \end{pmatrix}$$

$$y \to 1, \quad 1, \quad \begin{pmatrix} \zeta & 0 \\ 0 & 1/\zeta \end{pmatrix},$$

where $\zeta = e^{\frac{2\pi i}{3}}$. The irreducible representations of C_2 are given by

$$z \to 1, \quad -1.$$

Hence the irreducible representations of $D_3 \times C_2$ are given by

$$x \times 1 \to 1, \quad -1, \quad 1, \quad -1, \quad \begin{pmatrix} 0 & 1 \\ 1 & 0 \end{pmatrix}, \quad \begin{pmatrix} 0 & 1 \\ 1 & 0 \end{pmatrix}$$

$$y \times 1 \to 1, \quad 1, \quad 1, \quad 1, \quad \begin{pmatrix} \zeta & 0 \\ 0 & 1/\zeta \end{pmatrix}, \quad \begin{pmatrix} \zeta & 0 \\ 0 & 1/\zeta \end{pmatrix}$$

$$1 \times z \to 1, \quad 1, \quad -1, \quad -1, \quad I, \quad -I.$$

(b) $D_3 \times D_3$ has 9 classes and irreducible representations of degrees 1, 1, 1, 1, 2, 2, 2, 2, 4. The one of degree 4 may be given by

$$x \times 1 \to \begin{bmatrix} 0 & 0 & 1 & 0 \\ 0 & 0 & 0 & 1 \\ 1 & 0 & 0 & 0 \\ 0 & 1 & 0 & 0 \end{bmatrix}, \quad y \times 1 \to \begin{bmatrix} \zeta & 0 & 0 & 0 \\ 0 & \zeta & 0 & 0 \\ 0 & 0 & 1/\zeta & 0 \\ 0 & 0 & 0 & 1/\zeta \end{bmatrix},$$

$$1 \times x \to \begin{bmatrix} 0 & 1 & 0 & 0 \\ 1 & 0 & 0 & 0 \\ 0 & 0 & 0 & 1 \\ 0 & 0 & 1 & 0 \end{bmatrix}, \quad 1 \times y \to \begin{bmatrix} \zeta & 0 & 0 & 0 \\ 0 & 1/\zeta & 0 & 0 \\ 0 & 0 & \zeta & 0 \\ 0 & 0 & 0 & 1/\zeta \end{bmatrix}.$$

Since x and y have relatively prime periods, $x \times y$ and $y \times x$ will generate $D_3 \times D_3$. Thus $x \times 1 = (x \times y)^3$, $y \times 1 = (y \times x)^4$, etc.

Chapter VII

Bounded Representations and Weyl's Theorem

22 Hermitian and Bounded Representations

We now consider representations which are more general than representations of a finite group.

A representation $\alpha = \{A(x) : x \in G\}$ is said to be *hermitian* if a hermitian positive definite form

$$q(\tau) = \tau^* Q \tau = \sum_{i,j} q_{ij} \bar{\tau}_i \tau_j$$

exists such that

$$q(A(x)\tau) = q(\tau), \quad x \in G. \tag{49}$$

Alternatively, (49) says that

$$A(x)^* Q A(x) = Q, \quad x \in G. \tag{50}$$

If α is hermitian, then so is any representation equivalent to α. For if $\beta = \{B(x) : x \in G\}$ satisfies $\beta = S^{-1}\alpha S$, then $B(x) = S^{-1}A(x)S$ and (50) implies that

$$B(x)^* (S^* Q S) B(x) = S^* Q S, \quad x \in G.$$

Furthermore we have:

Theorem 27. *A reducible hermitian representation is fully reducible.*

Proof. We may suppose the given representation α to be reduced, so that

$$\alpha = \left\{ M(x) = \begin{bmatrix} A(x) & 0 \\ C(x) & D(x) \end{bmatrix} : x \in G \right\},$$

with corresponding form Q. Write $Q = T^*T$, T lower triangular. Then the representation $T\alpha T^{-1}$ is also reduced and has corresponding form

$$(T^{-1})^* Q T^{-1} = I.$$

Write

$$T\alpha T^{-1} = \left\{ M_1(x) = \begin{bmatrix} A_1(x) & 0 \\ C_1(x) & D_1(x) \end{bmatrix} : x \in G \right\}.$$

Then the matrices $M_1(x)$ are unitary. Thus

$$\begin{bmatrix} A_1(x) & 0 \\ C_1(x) & D_1(x) \end{bmatrix} \begin{bmatrix} A_1(x)^* & C_1(x)^* \\ 0 & D_1(x)^* \end{bmatrix} = I,$$

$$A_1(x)A_1(x)^* = I, \quad A_1(x)C_1(x)^* = 0,$$

which implies that $C_1(x) = 0$. Hence the representation $T\alpha T^{-1}$ is fully reduced, and the proof of the theorem is concluded. $\qquad\square$

Notice that this theorem provides an alternative approach to the study of finite groups, since any representation of a finite group is hermitian. For if $\alpha = \{A(x) : x \in G\}$ is a representation of a finite group G, then it possesses the associated form

$$Q = \sum_{x \in G} A(x)^* A(x).$$

The representation $\alpha = \{A(x) : x \in G\}$ is said to be *bounded* if a positive constant c exists such that

$$|a_{ij}(x)| < c, \quad x \in G.$$

The most important theorem about such representations is Weyl's theorem:

Theorem 28 (Weyl). *Every bounded representation is hermitian, and hence fully reducible.*

The proof of this theorem necessitates the introduction of some nonalgebraic concepts.

We first prove the following lemmas:

Lemma 13. *Let* $\alpha = \{A(x) : x \in G\}$ *be a bounded representation of degree* n. *Then* $|det(A(x))| = 1$, $x \in G$.

Proof. We have that for some suitable positive constant c,

$$|a_{ij}(x)| < c, \quad x \in G.$$

Then Hadamard's inequality implies that

$$|\det A(x)| < n^{\frac{n}{2}} c^n, \quad x \in G,$$

so that the determinants of the matrices of α are also uniformly bounded. Suppose that for some $x \in G$,

$$|\det(A(x))| = \rho \neq 1.$$

Then $|\det(A(x^p))| = \rho^p$ may be made arbitrarily large by choosing the integer p appropriately. This is a contradiction, and concludes the proof of the lemma. □

Lemma 14. *Let* P *be a subset of* R_n *(euclidean* n-*dimensional space), and* K(P) *its convex hull. Let* T *be any linear transformation of* R_n *into itself. Then*

$$T(K(P)) \subset K(T(P)).$$

Furthermore if T *is nonsingular and satisfies* T(P) = P, *then*

$$T(K(P)) = K(P).$$

Proof. Suppose that $\tau \in K(P)$. Then there are points $p_1, p_2, \ldots, p_t \in P$ and numbers $a_1, a_2, \ldots, a_t \geq 0$ such that $\tau = a_1 p_1 + a_2 p_2 + \cdots + a_t p_t$, $a_1 + a_2 + \cdots + a_t = 1$. Thus $T(\tau) = a_1 T(p_1) + a_2 T(p_2) + \cdots + a_t T(p_t)$, so that $T(\tau) \in K(T(P))$. This proves the first part of the lemma. To prove the second part, notice that the relationship $T(P) = P$ yields $T(K(P)) \subset K(P)$, and that the relationship $T^{-1}(P) = P$ yields $T^{-1}(K(P)) \subset K(P)$, $K(P) \subset T(K(P))$. This completes the proof. □

We now go on with the proof of Weyl's theorem. Our problem is the construction of the form $q(\tau)$. We regard the matrices $A(x)$ of α as linear transformations acting on the vectors τ of n complex variables according to the law $\tau' = A(x)\tau$. If we write τ, τ' and $A(x)$ in terms of their real and imaginary parts we have

$$\tau = u + iv, \quad \tau', = u' + iv', \quad A(x) = C(x) + iD(x),$$

and the complex transformation $\tau' = A(x)\tau$ is equivalent to the real transformation

$$u' = C(x)u - D(x)v, \qquad\qquad (51)$$
$$v' = D(x)u + C(x)v.$$

Put

$$\tilde{A}(x) = \begin{bmatrix} C(x) & -D(x) \\ D(x) & C(x) \end{bmatrix}, \quad x \in G.$$

Then the matrices $\tilde{A}(x)$ are real, and determine a representation $\tilde{\alpha} = \{\tilde{A}(x) : x \in G\}$ of G of degree $2n$. This is most easily seen by noticing that

$$\begin{bmatrix} C(x) & -D(x) \\ D(x) & C(x) \end{bmatrix} = \begin{bmatrix} I & iI \\ I & -iI \end{bmatrix}^{-1} \begin{bmatrix} A(x) & 0 \\ 0 & \overline{A(x)} \end{bmatrix} \begin{bmatrix} I & iI \\ I & -iI \end{bmatrix}.$$

Thus $\tilde{\alpha} \sim \alpha + \bar{\alpha}$ (and hence is reducible). Furthermore,

$$\det \tilde{A}(x) = |\det A(x)|^2 = 1,$$

by lemma 13.

We now consider the elements of $\tilde{\alpha}$ as linear transformations in euclidean $2n$-dimensional space R_{2n} acting on the unit sphere

$$E: \quad \sum_{k=1}^{n} (u_k^2 + v_k^2) \leqq 1,$$

where $u = (u_1, u_2, \ldots, u_n)^T$, $v = (v_1, v_2, \ldots, v_n)^T$. Denote the transformed region $\tilde{A}(x)(E)$ by E_x. Then E_x is a $2n$-dimensional ellipsoid, and the union

$$F = \bigcup_{x \in G} E_x$$

is a bounded point set of R_{2n}. Furthermore F is invariant with respect to every linear transformation $\tilde{A}(x)$.

Let K denote the convex hull of F. Then lemma 14 implies that K is also invariant with respect to every linear transformation $\tilde{A}(x)$.

The region K may be used as a region of integration, since it is a bounded convex subset of R_{2n}. Let $w_1, w_2, \ldots, w_n$ be n complex parameters, and define

$$q(w) = q(w_1, w_2, \ldots, w_n) = \int_K |w_1\tau_1 + w_2\tau_2 + \cdots + w_n\tau_n|^2 dV$$

$$= \int_K |(w, \tau)|^2 dV,$$

where $\tau = u + iv$ and dV is the volume element

$$dudv = du_1 du_2 \cdots du_n dv_1 dv_2 \cdots dv_n.$$

Clearly, $q(w)$ is a hermitian positive definite form in the n complex parameters $w_1, w_2, \ldots, w_n$. First make the transformation $w = A(x)^T \xi$. Then

$$q(A(x)^T \xi) = \int_K |(A(x)^T \xi, \tau)|^2 dV = \int_K |(\xi, A(x)\tau)|^2 dV.$$

Now make the complex transformation $\tau' = A(x)\tau$, corresponding to the real transformation (51). Then K is unchanged, the Jacobian of the transformation is 1, and we obtain

$$q(A(x)^T \xi) = \int_K |(\xi, \tau')|^2 dV',$$

where $dV' = du'dv' = du_1' du_2' \cdots du_n' dv_1' dv_2' \cdots dv_n'$. It follows that

$$q(A(x)^T \xi) = q(\xi),$$

and so the associated matrix Q satisfies

$$(A(x)^T)^* Q A(x)^T = Q, \quad x \in G.$$

Hence

$$A(x)^* Q_1 A(x) = Q_1, \quad x \in G,$$

where $Q_1 = (Q^{-1})^T$. But Q is hermitian positive definite if and only if Q_1 is. This completes the proof. $\qquad \square$

Appendix A
The Elements of the Theory of Algebraic Numbers

1 Fields. Polynomials Over Fields. Irreducibility

In this appendix a field will mean a subfield of the complex numbers; that is, a system of real or complex numbers which contains more than one number and is closed with respect to addition, subtraction, multiplication, and division, division by 0 being excluded.

Examples of fields are the rational numbers, the real numbers, and the complex numbers.

Every field contains a nonzero number α, and hence $\alpha/\alpha = 1$, $1+1 = 2$, $1 - 1 = 0$, and so on, thus all integers, and so all quotients of integers; that is, all rational numbers. The field of rational numbers will be denoted by R, and is contained in every field.

Let k be an arbitrary field. A polynomial whose coefficients belong to k will be referred to as a polynomial over k.

If $f(x), g(x)$ are polynomials, and $g(x)$ is of degree 1 at least, then polynomials $q(x), r(x)$ are uniquely determined such that

$$f(x) = q(x)g(x) + r(x), \tag{1}$$

and either $r(x)$ is identically 0 or the degree of $r(x)$ is less than the degree of $g(x)$. $r(x)$ is the remainder of $f(x)$ modulo $g(x)$. Furthermore, if $f(x)$, $g(x)$ are polynomials over k, then so are $q(x)$, $r(x)$. If $r(x)$ is identically 0, we say that $g(x)$ divides $f(x)$, or that $f(x)$ is divisible by $g(x)$, and we write $g(x)|f(x)$.

If in (1) the degree m of $f(x)$ is smaller than the degree n of $g(x)$, then $g(x) = 0$ and $r(x) = f(x)$. If however $m \geq n$, then the degree of $q(x)$ is $m - n$, $q(x)$ is not 0, and the degree of $r(x)$ is $< n$. If two polynomials $f(x)$, $g(x)$ divide one another, then their quotient is a constant. The constants are trivial divisors of every polynomial. A polynomial of degree 1 has only trivial divisors. The fundamental theorem of algebra tells us that every polynomial $f(x)$ of degree n can be written in exactly one way as a product of linear factors,

$$f(x) = c(x - \alpha_1)(x - \alpha_2) \cdots (x - \alpha_n),$$

where c is a nonzero constant and the α's are (not necessarily different) complex numbers. We see then that for polynomials over the field C of complex numbers the elements of C correspond to the units ± 1 and the linear polynomials to the primes of ordinary number theory.

The situation changes with polynomials over some given field k. We say that a polynomial $f(x)$ over k is *irreducible over* k, if $f(x)$ cannot be expressed as the product of two nonconstant polynomials over k.

An important theorem for polynomials over k is the following:

Theorem 1. *Two arbitrary nonzero polynomials* $f_1(x)$, $f_2(x)$ *over* k *possess a uniquely determined greatest common divisor* $d(x)$. *That is, there is a polynomial* $d(x)$ *with leading coefficient* 1 *such that*

$$d(x)|f_1(x), \quad d(x)|f_2(x)$$

and such that every dividing both $f_1(x)$ *and* $f_2(x)$ *also divides* $d(x)$. *Furthermore,* $d(x)$ *is representable in the form*

$$d(x) = g_1(x)f_1(x) + g_2(x)f_2(x), \tag{2}$$

where $g_1(x)$, $g_2(x)$ *are polynomials over* k, *and so* $d(x)$ *is itself a polynomial over* k.

Proof. Among the polynomials

$$L(x) = u_1(x)f_1(x) + u_2(x)f_2(x)$$

choose one of least degree and leading coefficient 1. Call this $d(x)$. Then (2) is satisfied. If $d(x)$ is of degree 0, then it is 1 and so divides both $f_1(x)$ and $f_2(x)$. If it is of higher degree, it must divide $f_1(x)$; for if $r(x)$ is the remainder of $f_1(x)$

modulo $d(x)$, then

$$f_1(x) = q(x)d(x) + r(x),$$
$$r(x) = f_1(x) - q(x)d(x)$$
$$= f_1(x) - q(x)\{g_1(x)f_1(x) + g_2(x)f_2(x)\}$$
$$= \{1 - q(x)g_1(x)\}f_1(x) + \{-q(x)g_2(x)\}f_2(x).$$

Thus $r(x)$ is one of the polynomials $L(x)$ whose degree is less than that of $f(x)$, and so must be 0. Hence $d(x)|f_1(x)$, and similarly, $d(x)|f_2(x)$. But (2) implies that every polynomial dividing both $f_1(x)$ and $f_2(x)$ divides $d(x)$.

Let $d_0(x)$ be some other polynomial with the properties of $d(x)$. Then $d_0(x)|d(x), d(x)|d_0(x)$, and so $d_0(x) = cd(x)$, for some constant factor c. Since the leading coefficients of $d_0(x)$ and $d(x)$ are 1, c must be 1. That is, $d_0(x) = d(x)$ and $d(x)$ is unique. This completes the proof. □

We write $(f_1(x), f_2(x)) = d(x)$ and say that $f_1(x), f_2(x)$ are *relatively prime*, if $d(x) = 1$.

A consequence of theorem 1 is

Theorem 2. *If* f(x) *is irreducible over* k *and if* g(x) *is a polynomial over* k *such that* f(x) *and* g(x) *have a zero* α *in common, then* f(x)|g(x).

Proof. Since $f(x)$ and $g(x)$ are both divisible by $x - \alpha, d(x) = (f(x), g(x)) > 1$. But $d(x)|f(x)$, and $f(x)$ is irreducible over k, so $d(x) = cf(x)$. Since $d(x)|g(x)$ as well, $f(x)|g(x)$. □

A conclusion we may draw from this theorem is that if $f(x)$ is irreducible over k, the zeros of $f(x)$ are distinct; for if not, $f(x)$ and $f'(x)$ (which is also in k) would have a zero in common, and so $f(x)|f'(x)$, which is impossible, since $f'(x)$ is of lower degree than $f(x)$.

2 Algebraic Numbers Over k

Let θ be a root of a polynomial over k. Of all polynomials over k with leading coefficient 1 having θ as a root, there is one of least positive degree. This is necessarily irreducible over k, and so is completely determined by θ and k, in view of theorem 2.

The degree n of this polynomial is called the *degree of* θ *with respect to* k (or the relative degree of θ). The n (distinct) roots $\theta_1, \theta_2, \ldots, \theta_n$ of this polynomial are called the *conjugates of* θ *with respect to* k (or the relative conjugates of θ). Such a number θ is called an *algebraic number over* k. If $k = R$, θ is simply called an *algebraic number*.

The numbers of k are all numbers of relative degree 1. For later purposes we will need the theorem on symmetric functions, which we put in the following form:

Let $\alpha_1, \alpha_2, \ldots, \alpha_n$ be n independent variables, and let $f_1, f_2, \ldots, f_n$ be the n elementary symmetric functions of $\alpha_1, \alpha_2, \ldots, \alpha_n$; i.e.,

$$(x - \alpha_1)(x - \alpha_2) \cdots (x - \alpha_n) = x^n - f_1 x^{n-1} + f_2 x^{n-2} - \cdots + (-1)^n f_n.$$

Then every polynomial function of the α's symmetric in the α's can be expressed as a polynomial function of the f's. The coefficients of the polynomial in the f's can be determined by addition, subtraction, and multiplication from the coefficients of the polynomial in the α's. (For a proof see e.g., Van der Waerden, Moderne Algebra, Vol. **1**.)

Applying this theorem twice, we have:

Let $\beta_1, \beta_2, \ldots, \beta_m$ be another set of independent variables, and $\varphi_1, \varphi_2, \ldots, \varphi_m$ the elementary symmetric functions of the β's. Then every polynomial function of the α's and the β's which is symmetric in the α's and in the β's is expressible as a polynomial function of the f's and the φ's, the coefficients of which can be determined by addition, subtraction, and multiplication from the coefficients of the polynomial in the α's and in the β's.

We are now able to show:

Theorem 3. *If α, β are algebraic numbers over* k, *then so are* $\alpha + \beta$, $\alpha - \beta$, $\alpha \cdot \beta$, *and* α / β, *if* $\beta \neq 0$.

Proof. Let $\alpha_1, \alpha_2, \ldots, \alpha_n$ be the conjugates of α and $\beta_1, \beta_2, \ldots, \beta_m$ the conjugates of β with respect to k. Then the elementary symmetric functions of the α's and the elementary symmetric functions of the β's are in k.

The product

$$H(x) = \prod_{j=1}^{m} \prod_{i=1}^{n} (x - (\alpha_i + \beta_j))$$

is symmetric in the α's and in the β's, and so is a polynomial over k, in view of the theorem above. Among the roots of $H(x)$, however, is $\alpha + \beta$, which implies that $\alpha + \beta$ is an algebraic number over k. In similar fashion we find that $\alpha - \beta$, $\alpha \cdot \beta$ are also algebraic numbers over k.

This procedure however is not applicable to α / β, $\beta \neq 0$. It is sufficient to show that $1/\beta$ is an algebraic number over k, since $\alpha / \beta = \alpha \cdot 1/\beta$. But if β is a root of the polynomial $f(x)$ over k of degree n, then $1/\beta$ is a root of the polynomial $g(x) = x^n f(1/x)$, which is also a polynomial over k.   □

Theorem 4. *Let ω be a root of the polynomial*

$$\varphi(x) = x^m + \alpha x^{m-1} + \beta x^{m-2} + \cdots + \lambda,$$

the coefficients of which are algebraic numbers over k. *Then ω is also an algebraic number over* k.

Proof. If α_i runs over the conjugates of α, β_j over the conjugates of β, and so forth, then the polynomial

$$F(x) = \prod_{i,j,\ldots,s} (x^m + \alpha_i x^{m-1} + \beta_j x^{m-2} + \cdots + \lambda_s)$$

is a polynomial over k, in view of the theorem on symmetric functions. Since $\varphi(x)|F(x)$, $F(\omega) = 0$, and so ω is an algebraic number over k. □

We conclude from the above that the algebraic numbers over k form an *algebraically closed* field.

3   Algebraic Extensions of k

Every number θ which is algebraic over k gives rise to a field containing k— namely, the field of rational functions of θ, which we denote by $k(\theta)$. Similarly, $k(\alpha, \beta, \ldots, \lambda)$, where $\alpha, \beta, \ldots, \lambda$ are numbers which are algebraic over k is defined as the field of rational functions of $\alpha, \beta, \ldots, \lambda$. $k(\alpha, \beta, \ldots, \lambda)$ is a *finite algebraic extension of k*, $k(\theta)$ *a simple algebraic extension of k.*

Theorem 5. *Every finite algebraic extension of* k *is equivalent to a simple algebraic extension of* k.

Proof. It is sufficient to consider only the case when two numbers are adjoined to k. Let $\alpha_1, \alpha_2, \ldots, \alpha_n$ be the n conjugates of a number α_1 of relative degree n over k and $\beta_1, \beta_2, \ldots, \beta_m$ the m conjugates of a number β_1 of relative degree m over k.

Consider the numbers

$$\omega_{ij} = \alpha_i + u\beta_j, \quad 1 \le i \le n; \ 1 \le j \le m,$$

where u is any number of k. The equations

$$\omega_{ij} = \omega_{i'j'}, \quad 1 \le i, \ i' \le n; \ 1 \le j, \ j', \ \le m,$$

have only a finite number of solutions u. Since k contains an infinite number of elements, there is a u in k such that the mn numbers ω_{ij} are distinct. We shall show that $k(\alpha_1, \beta_1) = k(\omega_{11})$. Since $\omega_{11} \in k(\alpha_1, \beta_1)$, we have $k(\omega_{11}) \subset k(\alpha_1, \beta_1)$. To show that $k(\alpha_1, \beta_1) \subset k(\omega_{11})$ we need only show that $\alpha_1 \in k(\omega_{11})$, $\beta_1 \in k(\omega_{11})$. To this end, consider the Lagrange interpolation polynomials

$$f(x) = \sum_{i,j} \frac{\alpha_i}{x - \omega_{ij}} D(x), \quad g(x) = \sum_{i,j} \frac{\beta_j}{x - \omega_{ij}} D(x)$$

where $D(x) = \prod_{i,j}(x - \omega_{ij}) \cdot f(x)$ and $g(x)$ are clearly polynomials over k, in view of the theorem on symmetric functions, and setting $x = \omega_{11}$,

$$D'(\omega_{11})\alpha_1 = f(\omega_{11}), \quad D'(\omega_{11})\beta_1 = g(\omega_{11}).$$

Now $D'(x)$ is a polynomial over k, and $D'(\omega_{11}) \neq 0$, since the ω's are distinct. Hence $\alpha_1 = \frac{f(\omega_{11})}{D'(\omega_{11})}, \beta_1 = \frac{g(\omega_{11})}{D'(\omega_{11})}$, and so are in $k(\omega_{11})$. This completes the proof. $\square$

Let θ be an algebraic number of relative degree n over k. We have:

Theorem 6. *Every element α of* $k(\theta)$ *is representable in one and only one way as*

$$\alpha = r(\theta),$$

where $r(x)$ *is a polynomial over* k *of degree not exceeding* $n - 1$.

Proof. Suppose $\alpha = P(\theta)/Q(\theta)$, where $P(x)$, $Q(x)$ are polynomials over k and $Q(\theta) \neq 0$. Let $g(x)$ be the irreducible polynomial over k with leading coefficient 1 having α as a root. By theorem 2, $(g(x), Q(x)) = 1$, since otherwise $g(x)$ and $Q(x)$ would have a root in common, $g(x)$ would divide $Q(x)$, and $Q(\theta)$ would be 0. Hence there are polynomials $R(x)$ and $S(x)$ over k such that $R(x)g(x) + S(x)Q(x) = 1$.

Setting $x = \theta$, we find $S(\theta)Q(\theta) = 1$, or $S(\theta) = 1/Q(\theta)$. Thus $\alpha = P(\theta)S(\theta) = f(\theta)$ say, where $f(x)$ is a polynomial over k.

We can write

$$f(x) = q(x)g(x) + r(x),$$

where $r(x)$ is of degree not exceeding $n - 1$. Setting $x = \theta$, we have $f(\theta) = r(\theta)$. Hence α has a representation in the desired form. Such a representation is necessarily unique, since $\alpha = r(\theta) = r_0(\theta)$ implies that $r(\theta) - r_0(\theta) = 0$. But the polynomial $r(x) - r_0(x)$ is a polynomial over k of degree less than the degree of $g(x)$. This is impossible unless $r(x) - r_0(x)$ is identically zero. $\square$

We also have the following:

Theorem 7. *Every number $\alpha = r(\theta)$ of* $k(\theta)$ *is an algebraic number over* k *of relative degree* n *at most. The relative conjugates of α are the different numbers among* $r(\theta_i)$, $1 \leq i \leq n$. *Every conjugate of α occurs the same number of times among these numbers.*

Proof. That α is an algebraic number over k of degree n at most may be seen by considering the polynomial

$$F(x) = \prod_{i=1}^{n}(x - r(\theta_i)).$$

The coefficients of $F(x)$ are symmetric polynomials in the θ's, and hence are in k, by the theorem on symmetric functions. $F(x)$ is therefore a polynomial over k of degree n of which α is a root. Let $\varphi(x)$ be the polynomial of least positive degree over k and leading coefficient 1 of which α is a root. Then $\varphi(x) = \Pi(x - \alpha_i)$, where $\alpha_i = r(\theta_i)$ runs over the distinct numbers of the set $\{r(\theta_1), r(\theta_2), \ldots, r(\theta_n)\}$. This is seen by observing (a) that $\varphi(r(x))$ and $g(x)$ have the root θ in common, so that

$\varphi(r(x))$ is divisible by $g(x)$ and so vanishes for $x = \theta_i$, $1 \le i \le n$; i.e., $\varphi(x)$ vanishes for $x = \alpha_i$, $1 \le i \le n$; and (b) that $\varphi(x)$ must have distinct roots.

Let the highest power of $\varphi(x)$ dividing $F(x)$ be p: $F(x) = \{\varphi(x)\}^p q(x)$, where $\varphi(x)$ does not divide $q(x)$. $q(x)$ must be constant, since otherwise $q(x)$ and $\varphi(x)$ would have a root in common, and $\varphi(x)$ would divide $q(x)$. Comparing leading coefficients, we find that this constant is 1. Thus $F(x) = \{\varphi(x)\}^p$, and the theorem is proved. □

We say that $k(\theta)$ *is of degree n over k.*

We notice that the relative degree of any number of $k(\theta)$ over k is a divisor of n, the degree of α being n/p.

If we fix an ordering of the θ's, and if $\alpha, \beta, \dots, \lambda$ are numbers of $k(\theta)$ the conjugates of which are $\alpha_i = \alpha(\theta_i)$, $\beta_i = \beta(\theta_i), \dots, \lambda_i = \lambda(\theta_i)$, then every rational equation $R(\alpha, \beta, \dots, \lambda) = 0$ with coefficients from k remains valid if the variables are replaced by their conjugates, attention being paid to the ordering.

4   Generators of Fields. Fundamental Systems. Subfields of $k(\theta)$

Theorem 8. *A number α of* k(θ) *belongs to* k *if and only if α is identical with its* n *conjugates. A number α of* k(θ) *is of degree* n *with respect to* k *if it is distinct from all its conjugates. The latter is a necessary and sufficient condition that α generate* k(θ).

Proof. If α is identical with its n conjugates, then the p of theorem 7 is n and α is of degree 1 over k; that is, α is in k. The converse is obvious. If α is distinct from all its conjugates, then all the conjugates must be distinct, since theorem 7 shows they occur with the same multiplicity, and so α is of degree n with respect to k. The converse here is clear also. □

Suppose that α generates $k(\theta)$, so that $k(\alpha) = k(\theta)$. Then the degree of α over k must be the degree of $k(\theta)$ over k; that is, n. Thus the conjugates of α are all distinct, and so α is certainly distinct from its conjugates. If α is distinct from its conjugates, then all the conjugates of α are distinct. We will show that θ can be expressed rationally in terms of α. Denoting the conjugates of α by $\alpha_i = r(\theta_i)$, $1 \le i \le n$, where $\theta = \theta_1, \alpha = \alpha_1$, we put

$$H(x) = \prod_{i=1}^{n}(x - \alpha_i) = \prod_{i=1}^{n}(x - r(\theta_i)),$$

$$K(x) = \sum_{i=1}^{n} \theta_i \frac{H(x)}{x - \alpha_i}.$$

$H(x)$, $K(x)$ are polynomials over k, and $K(\alpha_1) = \theta_1 H'(\alpha_1)$.

Since the α's are distinct, $H'(\alpha_1) \neq 0$, and we obtain $\theta_1 = K(\alpha_1)/H'(\alpha_1)$, a number in $k(\alpha)$.

We showed in theorem 6 that every number of $k(\theta)$ is a linear combination of $1, \theta, \ldots, \theta^{n-1}$ with coefficients from k. Generalizing this idea, we say:

The set of n numbers $\{\omega^{(1)}, \omega^{(2)}, \ldots, \omega^{(n)}\}$ is called a *fundamental system* of $k(\theta)$, if every number of $k(\theta)$ is a linear combination of these numbers with coefficients from k. $1, \theta, \ldots, \theta^{n-1}$ is thus a fundamental system of $k(\theta)$.

Theorem 9. *A necessary and sufficient condition that the numbers*

$$\omega^{(i)} = \sum_{j=1}^{n} c_{ij}\theta^{j-1}, \quad 1 \le i \le n, \; c_{ij} \; in \; k \tag{3}$$

be a fundamental system of $k(\theta)$ *is that* $\det(c_{ij}) \neq 0$.

Proof. Setting $C = (c_{ij})$, $r = (\omega^{(1)}, \omega^{(2)}, \ldots, \omega^{(n)})$, and $s = (1, \theta, \ldots, 0^{n-1})$, (3) reads

$$r^T = Cs^T.$$

If $\det C \neq 0$, then C^{-1} exists and $s^T = C^{-1}r^T$. The elements of C^{-1} are clearly in k, and so $\omega^{(1)}, \omega^{(2)}, \ldots \omega^{(n)}$ is a fundamental system of k since $1, \theta, \ldots, \theta^{n-1}$ is. If on the other hand the ω's form a fundamental system of k, then there is a matrix A with elements in k such that $s^T = Ar^T = ACs^T$. Hence $(I - AC)s^T = 0$. But $1, \theta, \ldots, \theta^{n-1}$ are linearly independent over k, and so we have identically $I - AC = 0$. C is therefore nonsingular. $\qquad\square$

An easy consequence of theorem 9 is

Theorem 10. *A necessary and sufficient condition that* n *elements of* k *be a fundamental system of* k *is that they be linearly independent over* k.

Theorem 10 allows us to define alternatively the degree of $k(\theta)$ over k as the maximum number of linearly independent elements of $k(\theta)$ over k; in other words, if $k(\theta)$ is regarded as a vector space over k, as the dimension of $k(\theta)$ over k.

Let us define for a fundamental system
$\omega^{(1)}, \omega^{(2)}, \ldots, \omega^{(n)} \Delta(\omega^{(1)}, \omega^{(2)}, \ldots, \omega^{(n)}) = \det(\omega_j^{(i)})$, the j denoting conjugates. From (3) we find easily

$$\Delta(\omega^{(1)}, \omega^{(2)}, \ldots, \omega^{(n)}) = \det C \, \Delta(1, \theta, \ldots, \theta^{n-1}).$$

$\Delta(1, \theta, \ldots, \theta^{n-1})$ is the well-known Vandermondian determinant

$$\begin{vmatrix} 1 & \theta_1 & \theta_1^2 & \cdots & \theta_1^{n-1} \\ 1 & \theta_2 & \theta_2^2 & \cdots & \theta_2^{n-1} \\ & & \cdots & & \\ 1 & \theta_n & \theta_n^2 & & \theta_n^{n-1} \end{vmatrix} = \prod_{1 \le i < j \le n} (\theta_j - \theta_i) \neq 0.$$

The square of this determinant is in k. $\Delta^2(\omega^{(1)}, \omega^{(2)}, \ldots, \omega^{(n)})$ is also in k, since $\det C$ is in k. $\Delta^2(\omega^{(1)}, \omega^{(2)}, \ldots, \omega^{(n)})$ is thus independent of the numbering of the conjugates.

We mention finally that if the degree of $k(\theta)$ over k denoted by $(k(\theta) : k)$, and if α is in $k(\theta)$, then

$$(k(\theta) : k) = (k(\theta) : k(\alpha))(k(\alpha) : k).$$

$k(\alpha)$ is a subfield of $k(\theta)$. If $k(\theta)$ is identical with $k(\theta_i)$ for all conjugates θ_i of θ over k, then $k(\theta)$ is said to be *normal* over k. $k(\alpha)$ is always a subfield of a normal extension of k — namely, the field obtained by adjoining to k all the conjugates of α with respect to k. This field is called the splitting field of α with respect to k.

5 Algebraic Integers

We now return to R (the rational field).

An algebraic number α of degree n over R is called an *algebraic integer* (more briefly, *integer*) if the coefficients of the irreducible equation with leading coefficient 1 for α are rational integers.

An algebraic integer which is rational is a rational integer. The conjugates of an algebraic integer are algebraic integers.

For the proof of the next theorem, we will require a lemma of Gauss.

Let $f(x)$ be a polynomial with rational integral coefficients. We write

$$f(x) \equiv 0 \bmod n$$

to mean that every coefficient of $f(x)$ is divisible by n. We also write $C(f)$ for the greatest common divisor of the coefficients of f. We easily verify that the ordinary laws of congruences are true in this context too.

Lemma 1. *Let* p *be a prime,* f(x), g(x) *polynomials with rational integral coefficients. Then*

$$f(x)g(x) \equiv 0 \bmod \text{p} \leftrightarrow f(x) \equiv 0 \bmod \text{p} \ or \ g(x) \equiv 0 \bmod \text{p}.$$

Proof. Suppose the lemma false. Then $f(x) \equiv f_1(x) \bmod p$, $g(x) \equiv g_1(x) \bmod p$, where $f_1(x)$, $g_1(x)$ are not zero, and no term in $f_1(x)$ or $g_1(x)$ is divisible by p. Thus

$$f_1(x)g_1(x) \equiv f(x)g(x) \equiv 0 \bmod p,$$

an impossibility, since the leading coefficient of $f_1(x)g_1(x)$ is certainly not divisible by p. $\qquad\square$

An easy consequence of lemma 1 is Gauss' lemma:

Gauss' Lemma. *Suppose that* C(f) = C(g) = 1. *Then* C(fg) = 1.

Theorem 11. *If α satisfies an equation with rational integral coefficients and leading coefficient 1, then α is an algebraic integer.*

Proof. Let $p(x), q(x)$ be polynomials with rational integral coefficients such that $p(x)$ has leading coefficient 1, $q(x)$ is irreducible, and $p(\alpha) = q(\alpha) = 0$. Then $q(x) | p(x)$, and we may set

$$p(x) = \frac{1}{c} q(x) Q(x),$$

where $Q(x)$ has rational integral coefficients and c is a rational integer. Let $a = C(q)$, $A = C(Q)$. Then $p(x) = \frac{aA}{c} q_0(x) Q_0(x)$, where $q_0(x) = \frac{q(x)}{a}$, $Q_0(x) = \frac{Q(x)}{A}$, and $C(q_0) = C(Q_0) = 1$. By Gauss' lemma, $C(q_0 Q_0) = 1$. But $C(p) = 1$, since $p(x)$ has leading coefficient 1. This implies that $\frac{aA}{c} = 1$, and $p(x) = q_0(x) Q_0(x)$. Since $p(x)$ has leading coefficient 1, so has $q_0(x)$. α is thus an algebraic integer. $\qquad\square$

By the methods of theorems 3, 4, we can now show:

Theorem 12. *The algebraic integers form a ring, which is algebraically closed.*

We make the further remark that if θ is an algebraic number, then there is a rational integer a_0 such that $a_0\theta$ is an algebraic integer. For if $a_0\theta^n + a_1\theta^{n-1} + \cdots + a_n = 0$ is the equation with rational integral coefficients which θ satisfies, then $a_0\theta$ satisfies $(a_0\theta)^n + a_1(a_0\theta)^{n-1} + \cdots + a_0^{n-1}a_n = 0$.

Let $K = R(\theta)$ be a field generated by the algebraic number θ of relative degree n. In virtue of the preceding remark, we may assume that θ is an algebraic integer. If α is in K, define

$$N(\alpha) = \alpha^{(1)}\alpha^{(2)} \cdots \alpha^{(n)}$$
$$S(\alpha) = \alpha^{(1)} + \alpha^{(2)} + \cdots + \alpha^{(n)},$$

superscripts denoting conjugates with respect to K. We note that $N(\alpha\beta) = N(\alpha)N(\beta)$, $S(\alpha + \beta) = S(\alpha) + S(\beta)$. $N(\alpha)$, $S(\alpha)$ are always rational numbers, and are rational integers if α is an integer. $N(\alpha) = 0$ only for $\alpha = 0$.

Theorem 13. *The algebraic integers of* K *form an infinite additive abelian group, possessing a basis of* n *elements. That is, there are* n *algebraic integers* $\omega_1, \omega_2, \ldots, \omega_n$ *in* K *such that every algebraic integer* α *in* K *has a unique representation in the form*

$$\alpha = x_1\omega_1 + x_2\omega_2 + \cdots + x_n\omega_n,$$

where the x*'s are rational integers. The* ω*'s are called a basis of* K.

Proof. That the algebraic integers of K form an infinite additive abelian group is clear. To prove the second part, we observe that every algebraic integer ρ in K has a unique representation in the form

$$\rho = c_0 + c_1\theta + \cdots + c_{n-1}\theta^{n-1},$$

where the c's are rational numbers. The c's are completely determined by the system

$$\begin{bmatrix} 1 & \theta^{(1)} & \cdots & \theta^{(1)n-1} \\ 1 & \theta^{(2)} & \cdots & \theta^{(2)n-1} \\ & \cdots & \\ 1 & \theta^{(n)} & \cdots & \theta^{(n)n-1} \end{bmatrix} \begin{bmatrix} c_0 \\ c_1 \\ \vdots \\ c_{n-1} \end{bmatrix} = \begin{bmatrix} \rho^{(1)} \\ \rho^{(2)} \\ \vdots \\ \rho^{(n)} \end{bmatrix},$$

superscripts denoting conjugates with respect to K. Solving this system, we find

$$\Delta c_k = \Delta_k, \quad 0 \le k \le n-1,$$

where $\Delta = \Delta(1, \theta, \ldots, \theta^{n-1}) \neq 0$ and Δ_k is a determinant the elements of which involve only nonnegative powers of the ρ's and the θ's. Δ_k is thus an algebraic integer.

From

$$c_k = \frac{\Delta_k}{\Delta} = \frac{\Delta\Delta_k}{\Delta^2},$$

we conclude that $\Delta\Delta_k = c_k\Delta^2$ is a rational integer, the right-hand side being rational, and the left-hand side integral. Hence $c_k = x_k/D$, where x_k is a rational integer and $D = |\Delta^2|$ is a rational integer which is *independent of* ρ.

We consider now the totality $\left\{x_0\frac{1}{D} + x_1\frac{\theta}{D} + \cdots + x_{n-1}\frac{\theta^{n-1}}{D}\right\}$, where the x's run over all rational integers. This is an additive abelian group with the finite basis $\{1/D, \theta/D, \ldots, \theta^{n-1}/D\}$. The previous discussion shows that the algebraic integers of K are a subgroup of this group, as a matter of fact of finite index D^n. Hence the group of algebraic integers of K also has a finite basis of not more than n elements. That there are precisely n elements in the basis may be seen by observing that the algebraic integers $1, \theta, \ldots, \theta^{n-1}$ of K are linearly independent.

The number

$$d = \begin{vmatrix} \omega_1^{(1)} & \omega_2^{(1)} & \cdots & \omega_n^{(1)} \\ \omega_1^{(2)} & \omega_2^{(2)} & \cdots & \omega_n^{(2)} \\ & \cdots & \\ \omega_1^{(n)} & \omega_2^{(n)} & \cdots & \omega_n^{(n)} \end{vmatrix}^2$$

is called the *discriminant* of K, and is independent of the particular basis chosen. Any n linearly independent algebraic integers of k making d minimal will do for a basis.

An algebraic integer ϵ is a *unit* if and only if $1/\epsilon$ is an algebraic integer. Alternatively, ϵ is a unit if and only if $N(\epsilon) = \pm 1$. Two algebraic integers whose quotient is a unit are said to be *associates*. □

6   Ideals

A system S of algebraic integers of a field $K = R(\theta)$ is called an *ideal in K* (or *ideal*) if for α, β in S and algebraic integers λ, μ in K, $\lambda\alpha + \mu\beta$ is in S.

Theorem 14. *Every ideal* S *can be put in the form* $S = \{x_1\alpha_1 + x_2\alpha_2 + \cdots + x_n\alpha_n\}$, *where the* x*'s range over all rational integers and the* α*'s are fixed elements of* S.

Proof. We assume that S does not consist of 0 alone. The elements of S form an additive group, which is a subgroup of the additive abelian group of the algebraic integers of K. This group has a finite basis of n elements, by theorem 13, and so S has a finite basis also. The number of elements in the basis is thus $\leq n$. But if $\alpha \neq 0$ is in S, then so are $\alpha, \theta\alpha, \ldots, \theta^{n-1}\alpha$, and these are linearly independent. The number of elements in the basis is thus precisely n.

Let $(\alpha_1, \alpha_2, \ldots, \alpha_r)$ denote the totality $\{\lambda_1\alpha_1 + \lambda_2\alpha_2 + \cdots + \lambda_r\alpha_r\}$, where the λ's run over all algebraic integers in K. Theorem 14 shows that any ideal $S = (\alpha_1, \alpha_2, \ldots, \alpha_r)$, where $r \leq n$. It may be shown that $r = 2$ always suffices. S is called *principal* if $S = (\alpha)$. The ideals $A = (\alpha_1, \alpha_2, \ldots, \alpha_r)$ and $B = (\beta_1, \beta_2, \ldots, \beta_s)$ are equal if and only if every α is in B and every β is in A. The principal ideals (α) and (β) are equal if and only if $\alpha|\beta$, $\beta|\alpha$; i.e., if and only if α and β are associates.   □

The product of the ideals A and B is defined as the ideal

$$(\alpha_1\beta_1, \ldots \alpha_i\beta_j, \ldots, \alpha_r\beta_s).$$

It is easy to show that this definition is independent of the particular basis chosen. We can easily verify that multiplication is associative and commutative. We define powers of an ideal in the usual way, and divisibility likewise. We note that the principal ideal (α) divides the principal ideal (β) if and only if α divides β. The totality of algebraic integers of K is the ideal (1). The ideal consisting of 0 along is (0). An ideal P is called a prime ideal if the only divisors of P are (1) and P.

Theorem 15. *For any ideal* A *there is an ideal* B *such that* AB *is principal.*

The proof requires certain preliminaries.

Lemma 2. *Let*

$$f(x) = \delta_0 + \delta_1 x + \cdots + \delta_m x^m, \quad \delta_m \neq 0$$

be a polynomial with integral coefficients. Let ρ *be a root of* f(x). *Then* $\frac{f(x)}{x-\rho}$ *also has integral coefficients.*

Proof. If $m = 1$ then $\rho = -\frac{\delta_0}{\delta_1}$,

$$\frac{f(x)}{x - \rho} = \frac{\delta_0 + \delta_1 x}{x + \frac{\delta_0}{\delta_1}} = \delta_1.$$

The lemma is thus true for $m = 1$.

Suppose the lemma proved for all polynomials of degree $\leq m - 1$. Consider

$$\varphi(x) = f(x) - \delta_m x^{m-1}(x - \rho).$$

Then $\deg \varphi(x) \leq m - 1$, and $\varphi(x)$ has integral coefficients, since $\delta_m \rho$ is an integer. Furthermore ρ is a root of $\varphi(x)$. By the induction hypothesis, $\frac{\varphi(x)}{x-\rho}$ has integral coefficients. Since $\frac{\varphi(x)}{x-\rho} = \frac{f(x)}{x-\rho} - \delta_m x^{m-1}$, it follows that $\frac{f(x)}{x-\rho}$ also has integral coefficients, and the proof of the lemma is complete. $\qquad\square$

Lemma 3. *If* $f(x)$ *given in lemma 2 above satisfies*

$$f(x) = \delta_m(x - \rho_1)(x - \rho_2) \cdots (x - \rho_m),$$

then $\delta_m \rho_1 \rho_2 \cdots \rho_k$ *is an integer for* $1 \leq k \leq m$.

Proof. By repeated application of lemma 2 the coefficients of

$$f_k(x) = \frac{f(x)}{(x - \rho_{k+1}) \cdots (x - \rho_m)} = \delta_m(x - \rho_1) \cdots (x - \rho_k)$$

must all be integers; and since

$$\delta_m \rho_1 \rho_1 \cdots \rho_k = (-1)^k f_k(0),$$

the lemma follows. $\qquad\square$

Lemma 4. *Let*

$$g(x) = \alpha_1 + \alpha_2 x + \cdots + \alpha_r x^{r-1}$$
$$h(x) = \beta_1 + \beta_2 x + \cdots + \beta_s x^{s-1}$$

where the α's and the β's are integral. Suppose that all the coefficients of $g(x)h(x)$ *are divisible by the algebraic integer v. Then all the products*

$$\alpha_i \beta_j, \quad 1 \leq i \leq r, \ 1 \leq j \leq s,$$

are divisible by v.

Proof. Suppose that

$$g(x) = \alpha_r(x - \rho_1) \cdots (x - \rho_{r-1}),$$
$$h(x) = \beta_s(x - \sigma_1) \cdots (x - \sigma_{s-1}).$$

By assumption the coefficients of

$$\frac{g(x)h(x)}{\nu} = \frac{\alpha_r \beta_s}{\nu}(x - \rho_1) \cdots (x - \sigma_{s-1})$$

are integral. Thus by lemma 3 every product

$$\frac{\alpha_r \beta_s}{\nu} \rho_{n_1} \rho_{n_2} \cdots \rho_{n_i} \sigma_{m_1} \sigma_{m_2} \cdots \sigma_{m_k}$$

is integral, where the n_ρ are arbitrary distinct integers between 1 and $r-1$ (inclusive) and the m_q arbitrary distinct integers between 1 and $s-1$ (inclusive). But $\frac{\alpha_i}{\alpha_r}, \frac{\beta_k}{\beta_s}$ are elementary symmetric functions of the ρ's and the σ's, respectively; say $\frac{\alpha_i}{\alpha_r} = S_i, \frac{\beta_k}{\beta_s} = S_k$. Thus

$$\frac{\alpha_i \beta_k}{\alpha_r \beta_s} = S_i S_k,$$
$$\frac{\alpha_i \beta_k}{\nu} = \frac{\alpha_r \beta_s}{\nu} S_i S_r,$$

and thus is an integer. This completes the proof of the lemma.   □

We go on now to the proof of theorem 15. Let $A = (\alpha_1, \alpha_2, \ldots, a_r)$, where $a_r \neq 0$. With A we associate the polynomial $g(x) = \alpha_1 + \alpha_2 x + \cdots + \alpha_r x^{r-1}$. Let $h(x)$ be such that

$$g(x)h(x) = \prod_{i=1}^{n}(\alpha_1^{(i)} + \alpha_2^{(i)}x + \cdots + \alpha_r^{(i)}x^{r-1}),$$

superscripts denoting conjugates with respect to K. The coefficients of $g(x)h(x)$ are rational integers. Moreover, the coefficients of $h(x)$ must be in K, since the coefficients of $g(x)$ and of $g(x)h(x)$ are in K. Hence the coefficients of $h(x)$ are integers of K. We set $h(x) = \beta_1 + \beta_2 x + \cdots + \beta_s x^{s-1}$. β_s is not zero, and $s - 1 = (n-1)(r-1)$. Let

$$B = (\beta_1, \beta_2, \ldots, \beta_s).$$

We shall show that

$$AB = (a),$$

where a is the greatest common divisor of the coefficients of $g(x)h(x)$.

Since a is a linear combination with rational integral coefficients of the coefficients of $g(x)h(x)$, and since the coefficients of $g(x)h(x)$ are sums of terms $\alpha_i\beta_j$, it is clear that a is in AB. Conversely, since a divides all the coefficients of $g(x)h(x)$, a divides every term $\alpha_i\beta_j$, by lemma 4, and so every element of AB is in (a). Hence $AB = (a)$. This completes the proof.   □

Theorem 16. *If* A $\neq$ (0) *and* AB = AC, *then* B = C.

Proof. Let A^* be such that $AA^* = (\alpha)$. Then $AB = AC$ implies that $AA^*B = AA^*C$, or $(\alpha)B = (\alpha)C$. Thus $\alpha \times$ any element of B is $\alpha \times$ an element C, and vice versa. Hence $B = C$.   □

Theorem 17. C|A *if and only if* C $\supset$ A.

Proof. If $C|A$, there is an ideal B such that $A = BC$. Thus if α is in A,

$$\alpha = \left(\sum_i x_i\beta_i\right)\left(\sum_j y_j\gamma_j\right)$$

$$= \sum_j \left\{\sum_i x_iy_j\beta_i\right\}\gamma_j,$$

which is in C. Suppose now that $C \supset A$. Let C^* be such that $CC^* = (\gamma)$. Then, as is easily shown, $CC^* \supset AC^*, (\gamma) > AC^*$. Setting $AC^* = (\rho_1, \rho_2, \ldots)$ we see that $\rho_i = \gamma\beta_i$, so that $AC^* = (\gamma\beta_1, \gamma\beta_2, \ldots) = (\gamma)(\beta_1, \beta_2, \ldots) = (\gamma)B$, say. Hence $AC^* = CC^*B$, $A = BC$, that is, $C|A$.   □

We conclude from the above that $A|C$, $C|A$ if and only if $A = C$.

Theorem 18. *The two ideals* A = $(\alpha_1, \alpha_2, \ldots, \alpha_r)$ *and* B = $(\beta_1, \beta_2, \ldots, \beta_s)$ *which are not both* (0) *have a uniquely determined greatest common divisor* D = (A, B). D *is just the ideal* $(\alpha_1, \alpha_2, \ldots, \alpha_r, \beta_1, \beta_2, \ldots, \beta_s)$ *(or the smallest ideal containing both* A *and* B).

Proof. Since $D \supset A$, $D \supset B$, $D|A$ and $D|B$. Let D_0 be any other ideal such that $D_0|A$, $D_0|B$. Then $D_0 \supset A$, $D_0 \supset B$. But D is the smallest ideal containing A and B and so $D_0 \supset D$. That is, $D_0|D$.

If further D_0 is any other ideal with the properties of D, then $D_0|D$, $D|D_0$ and so $D_0 = D$.   □

An easy deduction is that

$$C(A, B) = (CA, CB).$$

We can now prove:

Theorem 19. *If* P *is a prime ideal such that* P|AB, *then* P|A *or* P|B.

Proof. Suppose that P does not divide B, so that $(P, B) = (1)$. Then $A = A(1) = A(P, B) = (AP, AB)$. But $P|AB$, and so $P|A$.   □

Theorem 20. *A positive rational integer* a *belongs to at most a finite number of ideals.*

Proof. Let $\omega_1, \omega_2, \ldots, \omega_n$ be a basis for (1), in the sense of theorem 13. Every integer α of (1) has a representation

$$\alpha = x_1\omega_1 + x_2\omega_2 + \cdots + x_n\omega_n,$$

where the x's are rational integers. We reduce the x's modulo a, obtaining

$$x_i = aq_i + r_i, \quad 0 \le r_i \le a - 1, \ 1 \le i \le n.$$

Thus

$$\alpha = \sum_{i=1}^{n} x_i\omega_i = \sum_{i=1}^{n}(aq_i + r_i)\omega_i = aQ + R,$$

say, where $Q = \sum_{i=1}^{n} q_i\omega_i$, $R = \sum_{i=1}^{n} r_i\omega_i$. Q is integral and R assumes at most a^n values. Let A be the ideal $(\alpha_1, \alpha_2, \ldots, \alpha_r)$, $r \le n$. Using the representation for α given above, we find

$$A = (Q_1a + R_1, Q_2a + R_2, \ldots, Q_ra + R_r).$$

Suppose now that a is in A. Then

$$A = (Q_1a + R_1, Q_2a + R_2, \ldots, Q_ra + R_r, a) = (R_1, R_2, \ldots, R_r, a).$$

Since there are no more than a^n choices for a particular R, the total number of ideals A does not exceed a^{nr}. Since $r \le n$, the number of ideals A to which a belongs is finite. □

Theorem 21. *Every ideal* A *has only a finite number of divisors.*

Proof. Let A^* be such that
$$AA^* = (a),$$

where a is a positive rational integer. If $B|A$, then $B|(a)$, since $A|(a)$. Hence a is in B. But a belongs to at most a finite number of ideals B, by theorem 20. Thus the number of divisors B of A is finite. □

Theorem 22. *If* A $\neq$ (0) *and* A $=$ BC, *where* C $\neq$ (1), *then* B *has fewer divisors than* A.

Proof. Every divisor of B is a divisor of A. A however is also a divisor of A, but not of B, since $C \neq (1)$. Hence B has fewer divisors than A. □

A consequence of theorems 21 and 22 is:

Theorem 23. *Every ideal* A $\neq$ (0) *is divisible by a prime ideal (thus there exists at least one prime ideal).*

Assembling the preceding theorems, we see that the unique factorization theorem holds true for ideals, the proof being identical with that given for the rational integers:

Theorem 24. *Every ideal* A $\neq$ (0) *can be expressed as the product of a finite number of prime ideals. Apart from order, the expression is unique.*

Appendix B
The Roots of Unity

Let n be a positive integer. A number θ is said to be an nth *root of unity* if $\theta^n = 1$, and is *primitive* if n is the least positive integer such that $\theta^n = 1$. Then the nth roots of unity are the roots of the polynomial $x^n - 1$, and the primitive nth roots of unity are the roots of $x^n - 1$ which are not roots of $(x - 1)(x^2 - 1) \cdots (x^{n-1} - 1)$. It follows that the primitive nth roots of unity are the roots of $(x^n - 1)/D(x)$, where $D(x)$ is the greatest common divisor of the polynomials $x^n - 1$ and $(x - 1)$ $(x^2 - 1) \cdots (x^{n-1} - 1)$.

By theorem 11 of appendix A, the nth roots of unity are algebraic integers. They consist of the numbers

$$\zeta^k, \ 1 \le k \le n, \ \zeta = e^{\frac{2\pi i}{n}}.$$

The primitive nth roots of unity are the numbers

$$\zeta^k, \ (k, n) = 1, \ 1 \le k \le n.$$

Put

$$\Phi_n(x) = \prod_{\substack{(k,n)=1 \\ 1 \le k \le n}} (x - \zeta^k).$$

Then $\Phi_n(x)$ is the *cyclotomic polynomial* of order n. By the remarks above, $\Phi_n(x)$ must have rational integral coefficients.

An exact expression for $\Phi_n(x)$ is easily obtained. The numbers $1, 2, \ldots, n$ may be broken up into sets S_d, where $d \mid n$ and S_d consists of the numbers k such that $(k, d) = 1, 1 \leq k \leq d$. Then $\frac{n}{d}k$ runs over the numbers $1, 2, \ldots, n$ once and once only as d runs over the divisors of n and k runs over S_d, and $\zeta^{\frac{nk}{d}}$ runs over the primitive dth roots of unity as k runs over S_d. It follows that

$$x^n - 1 = \prod_{d/n} \Phi_d(x);$$

and by the Möbius inversion formula,

$$\Phi_n(x) = \prod_{d/n}(x^{n/d} - 1)^{\mu(d)}.$$

Since the degree of $\Phi_n(x)$ is

$$\sum_{\substack{(k, n)=1 \\ 1 \leq k \leq n}} 1 = \varphi(n),$$

we have proved:

Theorem 1. *There are precisely* $\varphi(n)$ *primitive* nth *roots of unity, and these are the roots of the cyclotomic polynomial*

$$\Phi_n(x) = \prod_{\substack{(k, n)=1 \\ 1 \leq k \leq n}} (x - \zeta^k) = \prod_{d/n}(x^{n/d} - 1)^{\mu(d)},$$

which is monic and has rational integral coefficients. The primitive nth *roots of unity are thus algebraic integers of degree* $\leq \varphi(n)$.

Our next task is to show that $\Phi_n(x)$ is irreducible over the rationals. The proof we give is technically elementary in that it avoids the use of the Dirichlet theorem on primes in arithmetic progressions, and is reasonably simple. It is possible to give a much simpler proof if this theorem is used.

We first prove

Lemma 1. *Let* k, n *be arbitrary rational integers such that* $n > 0$, $(k, n) = 1$; *and let* c *be an arbitrary positive number. Then there always exists a solution of the congruence*

$$t \equiv k \ mod \ n$$

such that each prime divisor of t *is greater than* c.

Proof. Let $q_1, q_2, \ldots, q_r$ be the primes $\leq c$ which do not divide n, and put $q = q_1 q_2 \cdots q_r$. Since $(q, n) = 1$, an integer t may be determined (by the Chinese remainder theorem) such that

$$t \equiv k \bmod n,$$
$$t \equiv 1 \bmod q.$$

Then $(t, q_i) = 1$, and certainly $(t, n) = 1$ (since $(k, n) = 1$). It follows that no prime $\leq c$ can divide t. This completes the proof. $\qquad\square$

Now let $f(x)$ be the monic irreducible polynomial with rational integral coefficients of which ζ is a root. We will show that ζ^k is also a root of $f(x)$ for each k such that $(k, n) = 1$, $1 \leq k \leq n$. Since $f(x)|\Phi_n(x)$ (theorem 2 of appendix A) this will imply that in fact $f(x) = \Phi_n(x)$, which in turn will imply that $\Phi_n(x)$ is irreducible over the rationals.

Let λ be any nth root of unity, and let $\lambda_1 = \lambda, \lambda_2, \ldots, \lambda_h$ be the conjugates of λ. Then $\lambda_1, \lambda_2, \ldots, \lambda_h$ are all nth roots of unity. Thus there is a positive constant c such that $|f(\lambda_i)| \leq c$, $1 \leq i \leq h$ (for example, the sum of the absolute values of the coefficients of $f(x)$). It follows that if m is a rational integer $> c$ such that $f(\lambda)/m$ is an integer, then $f(\lambda) = 0$ (since $N(f(\lambda)/m)$ is a rational integer of absolute value < 1).

Now suppose that t is a positive integer such that $f(\zeta^t) = 0$, and let p be any rational prime $> c$. Then by an elementary multinomial identity,

$$f(\zeta^{pt}) = f(\zeta^t)^p \equiv 0 \bmod p,$$

and hence $f(\zeta^{pt})/p$ is an integer. By the remark above, $f(\zeta^{pt})$ must be zero. Since $f(\zeta) = 0$, we conclude by induction that $f(\zeta^t) = 0$ for all rational integers t such that each prime divisor of t is $> c$.

Now choose t by lemma 1 so that $t \equiv k \bmod n$ and each prime divisor of t is $> c$. Then $\zeta^k = \zeta^t$, $f(\zeta^k) = f(\zeta^t) = 0$. Thus we have proved.

Theorem 2. *The cyclotomic polynomial $\Phi_n(x)$ is irreducible over the rationals, and hence the primitive nth roots of unity are algebraic integers of degree $\varphi(n)$.*

We now go on to obtain some elementary bounds for the function $\varphi(n)$. Let

$$n = p_1^{e_1} p_2^{e_2} \cdots p_s^{e_s}, \quad p_1 < p_2 < \cdots < p_s,$$

be the prime power decomposition of n. Then as is well-known,

$$\varphi(n) = n \left(1 - \frac{1}{p_1}\right) \left(1 - \frac{1}{p_2}\right) \cdots \left(1 - \frac{1}{p_s}\right).$$

Now $p_1 \geq 2$, $p_2 \geq 3, \ldots, p_s \geq s + 1$. Thus

$$\varphi(n) \geq n \left(1 - \frac{1}{2}\right) \left(1 - \frac{1}{3}\right) \cdots \left(1 - \frac{1}{s+1}\right) = \frac{n}{s+1}. \tag{1}$$

Also,

$$n = p_1^{e_1} p_2^{e_2} \cdots p_s^{e_s} \geq 2^{e_1 + e_2 + \cdots + e_s} \geq 2^s,$$

so that

$$s \leq \log n / \log 2. \tag{2}$$

Combining (1) and (2), we obtain

Theorem 3. *The function $\varphi(n)$ satisfies*

$$\varphi(n) \geq \frac{n \log 2}{\log(2n)}. \tag{3}$$

This result provides an effective means for determining the values of n for which $\varphi(n)$ takes a given value. For example, (3) implies that if $\varphi(n) \leq 2$ then $n \leq 8$; and an examination of the actual values assumed by $\varphi(n)$ for $1 \leq n \leq 8$ shows that in fact n must be 1, 2, 3, 4, 6.

The bound is certainly not best possible. It can be shown that actually

$$\varphi(n) \geq c \frac{n}{\log \log n}$$

for some appropriate positive constant c.

References

(It goes without saying that this list is incomplete. It consists of those books, or papers which the writer considers useful references, or which were referred to in the text.)

[1] Boerner, H., Representations of groups, Amsterdam (1963).

[2] Brenner, J. L., Quelques groupes libres de matrices, C. R. Acad. Sci. Paris **241**, 1689–1691 (1955).

[3] Burnside, W., On group characteristics, Proc. London Math. Soc. **33**, 46–62 (1901).

[4] Burnside, W., On the conditions of reducibility of any group of linear substitutions, Proc. London Math. Soc. **3**, 430–434 (1905).

[5] Burnside, W., On criteria for the finiteness of the order of a group of linear substitutions, Proc. London Math. Soc. **3**, 435–440 (1905).

[6] Burnside, W., Theory of Groups of Finite Order, 2d ed. (Cambridge University Press, 1911).

[7] Burrow, M., Representation Theory of Finite Groups (Academic Press, New York, 1965).

[8] Coxeter, H. S. M. and Moser, W. O. J., Generators and Relations for Discrete Groups, Berlin (1957).

[9]   Curtis, C. W. and Reiner, I., Representation Theory of Finite Groups and
      Associative Algebras (Interscience Publishers, New York, 1962).

[10] Dade, E. C., The maximal finite groups of 4×4 integral matrices, Ill. J. Math.
      9, 99–122 (1965).

[11] Feit, W., Characters of Finite Groups (W. U. A. Benjamin, Inc., New York,
      1967).

[12] Fluch, W., Über die Nichtlinearität einer gewissen Gruppe, Acta Arith. **10**,
      329–332 (1964).

[13] Frobenius, G., Über Gruppencharaktere, Sitzgsber. preuss. Akad. Wiss.
      985–1021 (1896).

[14] Frobenius, G., Über die Darstellung der endlichen Gruppen durch lin-
      eare Substitutionen: I. Sitzgsber. preuss. Akad. Wiss. 994–1015 (1897);
      II. Sitzgsber. preuss. Akad. Wiss. 482–500 (1899).

[15] Frobenius, G. and Schur, I., Über die reelen Darstellungen der endlichen
      Gruppen, Sitzgsber. preuss. Akad. Wiss. 186–208 (1906).

[16] Frobenius, G., Über die Äquivalenz der Gruppen linearer Substitutionen,
      Sitzgsber. preuss. Akad. Wiss. 209–217 (1906).

[17] Goldberg, K. and Newman, M., Pairs of matrices of order 2 which generate
      free groups, Ill. J. Math. **1**, 446–448 (1957).

[18] Hall, M., The theory of groups (The Macmillan Co., New York, 1959).

[19] Hecke, E., Vorlesungen über die Theorie der algebraischen Zahlen, Leipzig
      (1954).

[20] Higman, G., A finitely generated infinite simple group, J. London Math. Soc.
      26, 61–64 (1951).

[21] Kurosh, A. G., The theory of groups (I and II) (Chelsea, New York, 1955).

[22] Lomont, J. S., Applications of finite groups (Academic Press, New York,
      1959).

[23] Murnaghan, F. D., The theory of group representations (The Johns Hopkins
      Press, Baltimore, Md., 1938).

[24] MacDuffee, C. C., The theory of matrices (Chelsea, New York, 1946).

[25] Marcus, M., Basic theorems in matrix theory, NBS–AMS 57 (1960).

[26] Marcus, M. and Minc, H., A survey of matrix theory and matrix inequalities
      (Allyn & Bacon, Boston, 1964).

[27] Schur, I., Die algebraischen Grundlagen der Darstellungstheorie der Gruppen, Zuricher Vorlesungen (1936).

[28] Schur, I., Uber die rationalen Darstellungen der aligemeinen linearen Gruppe, Sitzgsber. preuss. Akad. Wiss. 58–75 (1927).

[29] Schur, I., Arithmetische Untersuchungen über endliche Gruppen linearer Substitutionen, Sitzgsber. preuss. Akad. Wiss. 164–184 (1906).

[30] Speiser, A., Die Theorie der Gruppen von endlicher Ordnung, Chelsea, New York (1954).

[31] Smirnov, V. I., Linear algebra and group theory (McGraw-Hill Book Co., Inc., New York, 1961).

[32] Van der Waerden, B. L., Gruppen von linearen Transformationen, Berlin (1935).

[33] Weyl, H., The classical groups, their invariants and representations (Princeton University Press, 1939).

Index